What in the World is Psychotherapy?

What in the World is *psychotherapy*

WRITTEN AND EDITED BY
Dr Austin Mardon, Annilea Purser, Benjamin Turner, Keshikaa Suthaaharan, Paawan Virdi, Priyanshu Mahey, Ghulam Aisha, Nawshin Haq, Neha Saroya, Nazihah Alam, Poojitha Pai, Hayley Zhong, Juwairia Razvi

Golden Meteorite Press
2021

Copyright © 2019 by Austin Mardon
All rights reserved. This book or any portion thereof may not be reproduced or used in any manner whatsoever without the express written permission of the publisher except for the use of brief quotations in a book review or scholarly journal.
First Printing: 2021

Typeset and Cover Design Jenna Usselman

978-1-77369-253-1
Golden Meteorite Press
103 11919 82 St NW
Edmonton, AB T5B 2W3
www.goldenmeteoritepress.com

Contents

1
What is The History of Psychotherapy?

13
Sigmund Freud: Who was he?

23
What is Psychotherapy?

31
What are the different types of psychotherapy?

39
What are the cognitive/behavioural effects of receiving Psychotherapy?

47
Impact of Psychotherapy on the Quality of Life

57
When is Psychotherapy Needed?

69
Is Psychotherapy Effective?

79
What is Psychotherapy in the modern world?

83
What are skepticisms surrounding Psychotherapy?

91
Public Perception of Psychotherapy

99
Future perspectives of psychotherapy

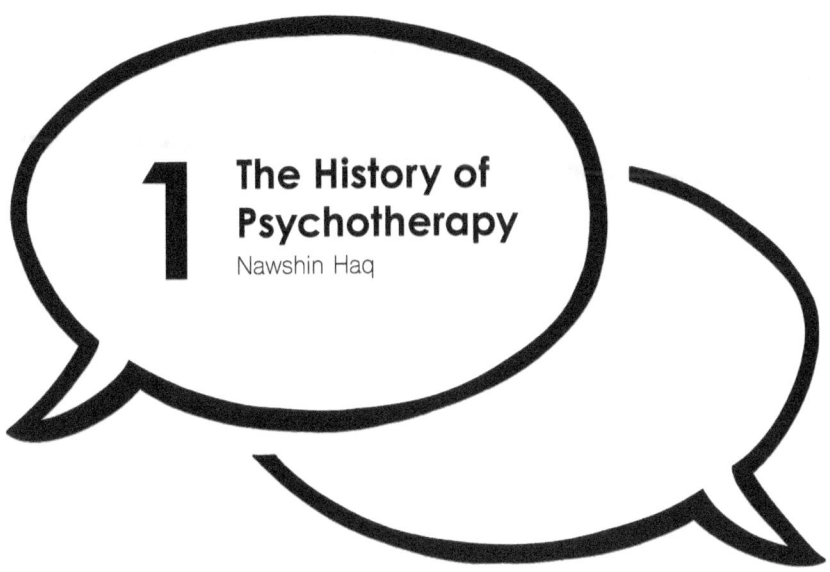

1 The History of Psychotherapy
Nawshin Haq

ABSTRACT

Psychotherapy is an essential field of treatment that has taken centuries to develop and centralize. Psychology has an extensive past of torturing mentally ill individuals; the initial advocates of psychotherapy were merely considerate individuals who actively protested for the appropriate treatment of psychologically distressed patients. Psychotherapy was disregarded as a field of research for many generations but eventually, due to relentless work from scientists and certain religious leaders led to the treatment method being accepted as a scientific technique. The origin and development of the field of study is deeply intertwined with religious and spiritual techniques. To understand the development of psychotherapy as a medical study, one must consider all original establishments and advocates of the field. This chapter will primarily focus on the leading figures in psychotherapy before Sigmund Freud and the development of psychoanalysis (which will be covered in the second chapter).

INTRODUCTION

History is a perplexing subject. Oftentimes humans ignore the significance of the history of a certain term or the creation of a certain method. Most times humans cannot trace, nor do they know the past of a specific concept. This is unfortunate since history is very important. The past assists us in building theories and plans for the future. Humans often make the same mistakes or have the same misconceptions repeatedly. Learning from the past can help us put a stop to this. Psychotherapy today may be known as the scientific treatment of mental disorders through psychological means (Norcross et al., 2011). However, this was not the case a century ago, in fact Psychotherapy was commonly associated with religious practices and conducted in spiritual settings (Norcross et al., 2011). This chapter focuses on the moral, spiritual, and religious background of Psychotherapy, and the gradual acceptance of Psychotherapy being used in a clinical context.

IMPORTANT CONSIDERATIONS
The history discussed in this chapter is from the W.E.I.R.D. (Western, Educated, Industrialized, Rich, and Democratic) perspective of the origin of Psychotherapy. There are, of course, various cultures in which Psychotherapy was and still is practised. For example, there is evidence that there are depictions of techniques in relation to modern Psychotherapy in Ancient Egyptian text (Hayen, 2016). It is depressing that scientists and historians sometimes do not make the effort to learn the holistic (or entire) nature of a method. Most of the information in this chapter originates from the book 'The History of Psychotherapy' which expands on mostly an American perspective of the method. Scientists are now making the effort, however, to try to figure out the origins of Psychotherapy in other cultures. One can only hope that the global scientific community takes into consideration the multicultural and multinational backgrounds of various modern concepts. The search for the true global history of Psychotherapy, however, is a work in progress. I will do my best to work with the information that is explained in Westernized papers and books, as I do not have access to other confirmed works of psychotherapy in other cultures.

There are significant gains in the Westernized research of Psychotherapy that has taken place after the 20th century. This chapter, however, will mostly explore the origins of Psychotherapy and how it prominently rose to be considered a treatment method by most physicians. Sigmund Freud's research and post-Freudian research on Psychotherapy will be explained in other chapters in further detail. It is important to note that Psychology and Psychotherapy has an extremely traumatic and distressing history. Due to mental illness being a serious topic and a chronic issue for many individuals, there are various upsetting situations that occurred to ill individuals before Psychology and psychological issues were considered seriously.

MORAL TREATMENT MOVEMENT

During the 1800sthe method of Psychotherapy started gaining momentum.any people were starting to withdraw from assumptions that people with mental illness were individuals possessed by malevolent spirits, thus using unjust methods of 'treating' these individuals through spiritual means (Norcross et al., 2011). The sorrowful yet undisputed truth was that patients with mental illnesses were often 'chained and kept away from human contact' (Norcross et al., 2011). The truly appalling methods of treatment included burning at the stake, flogging (spanking), keeping these individuals in truly horrifying conditions, etc (Ikiuga & Ciaravino, 2007). Moral treatment was an approach that was part of a larger chain of reform. It stems from the enlightenment period (1748-1751) and William Tuke (one of the founding figures) was influenced by the beliefs of the Quaker religion (Ikiuga & Ciaravino, 2007). It was established during the 1800s and primarily emphasized fair treatment for the mentally ill (Ikiuga & Ciaravino, 2007).

The movement was initially started by Philippe Pinel (1745-1826) who noticed the horrifying methods of 'treatment' that were used to cure the mentally ill (Ikiuga & Ciaravino, 2007). He was director for the Bicêtre and Salpêtrière asylums situated in Paris (Ikiuga & Ciaravino, 2007). During this time, he noticed the inhumane forms of care that were used on mentally ill patients. These treatments included bloodletting ('drawing a mentally ill patients' blood for therapeutic reasons' (Cohut, 2020)) and corporate punishment (physically abusing an individual due to misdemeanor) (Ikiuga & Ciaravino, 2007). He noticed how these methods were largely ineffective and decided to change the form of treatment bestowed upon patients afflicted with mental illness (Ikiuga & Ciaravino, 2007). It is important to note that he did not know nor properly understand the reasons for the mental illnesses inflicting these individuals; instead, he believed that mentally ill patients had a diminished morale and decreased sense of well-being (Ikiuga & Ciaravino, 2007). Thus, he proposed that they be treated with kindness and be allowed to live conceptually normal lives or perform daily activities (such as practicing hygiene, receiving regular labour, self care, etc.) (Ikiuga & Ciaravino, 2007). He believed that living a relatively normal lifestyle and being treated like a human being would help mentally ill individuals regain their sense of self (Ikiuga & Ciaravino, 2007). His methods of compassion and integration of new surroundings are still deployed in many modern-day theories of Psychotherapy (Ikiuga & Ciaravino, 2007).

While Phillipe Pinel was reforming Paris, another individual was moved by the tortuous treatment of the mentally ill. William Tuke (1732-1822) was a strong advocate of the Quaker religion and believed that those who are afflicted by mental illness should be treated with kindness (Ikiuga & Ciaravino, 2007). He sought out to create a retreat for the mentally ill where they could be safeand establish a life of peace (Ikiuga & Ciaravino, 2007). After establishing the retreat (York Retreat) he began trying to employ early intervention because he believed that coming forward with a method of treatment early in life could save an individual from a life of insanity (Ikiuga & Ciaravino, 2007). He also noticed that the patients responded to reinforcements thus, he started using these to attempt to change their behaviour (Ikiuga & Ciaravino, 2007).

Benjamin Rush (1746-1813) was a physician who sought to reform the treatment of the mentally ill after he noticed the various forms of torture inflicted on them as if they were criminals of an unforgivable crime (Ikiuga & Ciaravino, 2007). Thomas Scattergood (1748-1814) became a Minister and met William Tuke in England who introduced him to the York Retreat and the use of moral treatment methods (Ikiuga & Ciaravino, 2007). When he came back to America, he proposed the idea of a humane asylum. This was accepted by many (Ikiuga & Ciaravino, 2007). Thus, he and Benjamin Rush created the first asylum in the United States that used humane forms of cure to treat the mentally ill (Ikiuga & Ciaravino, 2007). This asylum was named the 'Friends Asylum' (Ikiuga & Ciaravino, 2007).

Dorothea Lynde Dix (1802-187) was an astounding individual with a difficult childhood.he became inspired by the works of Pinel, Tuke and Rush (Ikiuga & Ciaravino, 2007). She was voluntarily teaching Sunday school to women at the East Cambridge Jail in Massachusetts (Ikiuga & Ciaravino, 2007). There she noticed that criminals, prostitutes, mentally disabled and mentally ill individuals were all kept in the same quarters and treated in the same manner (Ikiuga & Ciaravino, 2007). When she asked why patients were kept in these staggering conditions, she was given the answer that her concerns were irrelevant (Ikiuga & Ciaravino, 2007). This inhumane and cruel treatment angered her, and she began a crusade across America for better care of the mentally ill (Ikiuga & Ciaravino, 2007). This crusade eventually led her to England where she met Queen Victoria and Pope Pius IX (Ikiuga & Ciaravino, 2007).

Thomas Kirkbade was a physician who became inspired by Benjamin Rush after reading his ideas in Rush on the mind, where he stated that humane treatment of the mentally ill could lead to their cure (Ikiuga & Ciaravino, 2007). As a resident at the Friends Asylum he employed various moral treatment methods on his patients (Ikiuga & Ciaravino, 2007). He advocated for Moral Treatment principles later on in his career as well (Ikiuga & Ciaravino, 2007). He would go on to become one of the founding members of the The Association of Medical Superintendents of American Institutions for the Insane, we now know this association as the American Psychiatric Association (APA) (Ikiuga & Ciaravino, 2007).

It is evident that these individuals from a variety of backgrounds felt an uncanny level of sympathy for the mistreated and tortured. When they recognized how dire the situation was, they sought to reform these horrendous conditions. The astonishing truth is that they had no idea that this movement would start a revolution (Ikiuga & Ciaravino, 2007). They only knew that these were patients and more importantly, human beings, who were being violently abused for a situation not within their control (Ikiuga & Ciaravino, 2007). The strong level of sympathetic understanding within the founders and advocates of the moral treatment displayed that human beings can be just as benevolent as they are cruel.

Although moral treatment originated from compassion and from a sympathetic standpoint the method was unfortunately abandoned by the 1890s (Norcross et al., 2011). Initially moral treatment was seen as a treatment method however, with time people started noticing that this type of method had no substantial effects or was not a suitable cure for chronic patients (Norcross et al., 2011). Due to this, mental institutions became a 'care' center in nature rather than a hospital (Norcross et al., 2011). Mental institutions were seen as lifelong sentences rather than a place where an ill individual could get treated (Norcross et al., 2011). Other treatment methods needed to be developed in order for mentally ill individuals to regain hope about their specific situations (Norcross et al., 2011).

Mesmerism/Animal Magnetism

Franz Anton Mesmer (1734-1815) was an Austrian Physician and a strong advocate for Mesmerism/Animal Magnetism (Norcross et al., 2011). He believed that there were invisible fluids within the patient's body and these fluids were magnetised, thus an imbalance in these fluids resulted in various neurological diseases (Norcross et al., 2011). Initially, he believed that placing magnets on an individual's body could restore the original balance in the bodily fluids (Norcross et al., 2011). He claimed that the addition of magnets was painful for the patient at first but soon became pleasurable as they were transported into a spellbound state (Norcross et al., 2011). As time passed, he developed the theory that many other objects (such as mirrors, music, and water) could have the same effect on the patient due to the 'magnetism' of these objects (Norcross et al., 2011).

Although this theory is quite different in nature in recent times, his theory is the reason that hypnosis (a certain state wherein an individual is susceptible to suggestion and apparently incapable of voluntary actions) grew in prominence (Norcross et al., 2011). He was also the first individual to try to explain the state of hypnosis using a scientific perspective (Norcross et al., 2011). The theory was originally 'debunked' in France after a group led by Benjamin Franklin studied the concept using empirical methods (Norcross et al., 2011). The committee were appalled by the use of animal magnetism and claimed it was similar to suggestion (leading the thoughts and behaviours of an individual usually using verbal instructions) (Norcross et al., 2011). His ideas were strongly rejected by the scientific community but increased in practice among other groups (Norcross et al., 2011). The method was brought to America by a French man named Charles Poyen (Norcross et al., 2011). He held large public demonstrations for the effects of animal magnetism and focused on the healing and spiritual effects of the technique (Norcross et al., 2011). This eventually led to 'Mesmerism' or hypnosis being largely practiced within spiritual groups and the occult (Norcross et al., 2011).

Mesmerism gave rise to a variety of other movements that eventually led to the increased popularity of Psychotherapy being practised by those who are not constantly intertwined with the medical community. These movements included the American Spirituality Movement, the Mind-Cure Movement and the Emmanuel Movement (Norcross et al., 2011).

AMERICAN SPIRITUALITY AND THE MIND-CURE MOVEMENT
The American Spirituality Movement was a sudden momentum when mediums and spiritual leaders grew exponentially in number (Norcross et al., 2011). The number of advocates for spirituality also grew to a substantial amount and at one point the movement had more than 11 million followers (Norcross et al., 2011). It began with the stories of the Fox sisters who convinced numerous people of the spirits within their house (Norcross et al., 2011). They claimed to hear a series of raps on their bedroom wall apparently made by an unseen presence (Norcross et al., 2011). Eventually, they confessed that the entire situation occurred due to them attempting to play a harmless April Fool's prank (Norcross et al., 2011). What they did not take into account was that they were the catalysts for the creation of American Spirituality (Norcross et al., 2011). During

this movement, there were mainly circumstances where mediums and spiritual individuals would communicate with 'spirits' (Norcross et al., 2011). These spiritual leaders would also at times provide counselling to grieving families and ill-people (Norcross et al., 2011).

The mind cure movement was started by Phineas Parkhurst Quimby (1802–1866) who initially practised Mesmerism and embarked on a journey to cure individuals through this method (Norcross et al., 2011). He developed the theory that Mesmerism was not the cure rather it was mental persuasion that healed these individuals (Norcross et al., 2011). He believed that the mind could heal the body and even stated that he was cured of semi-invalidism (semi-physical disability) through this method (Norcross et al., 2011). He eventually claimed he cured more than 12,000 individuals using his method (Norcross et al., 2011). One of these individuals was Mary Baker Eddy (1821–1910) who was adamant that there was a mental cure to all diseases and that mental therapeutics cured bodily ailments (Norcross et al., 2011). The mind-cure movement, although popular, was largely dismissed in the scientific community (Norcross et al., 2011). This was not the case with the Emmanuel Movement which not only put Psychotherapy in an enhanced position within the scientific community, but it also allowed physicians to gain the control of this movement within scientific grounds (Norcross et al., 2011).

EMMANUEL MOVEMENT

It is important to note that before the Emmanuel Movement Psychotherapy was not acknowledged as a scientific theory and not accredited to be rooted in any scientific knowledge (Norcross et al., 2011). Numerous physicians scoffed at those who practiced Psychotherapy and merely stated that Psychotherapy was a spiritual method (Norcross et al., 2011). They predominantly believed that the 'mind' did not have many effects on the body and thatmost research focused on the physiological aspects of diseases (Norcross et al., 2011). Therefore, without the Emmanuel Movement the state of Psychotherapy and the advancement of research in this field would not have occurred (Norcross et al., 2011). Practising Psychotherapy would have mainly remained as direct opposition to scientific methods of treatment (Norcross et al., 2011).

The Emmanuel Movement was started by Elwood Worcester (1862-1940). He earned his doctorate in Psychology and Philosophy from Leipzig while working with Wilhelm Wundt (1832-1920) (prominent figure in establishing Psychology as a scientific community) before becoming the minister of the Emmanuel Church. He strived to combine his scientific and religious knowledge to create a program that could offer help to individuals with nervous disorders (Norcross et al., 2011). In 1906, he established a program where individuals could come to the Church to watch lectures given by scientifically based figures (such as physicians), receive psycho-therapeutic help from the minister and free medical examinations (Norcross et al., 2011). The implementation of this inspiring program led to the acknowledgement of Psychotherapy as a treatment method by the medical community (Norcross et al., 2011). The program gained national popularity and attracted the attention of the general public as it attempted to cure matters of the mind rather than hoping they would simply alleviate through physical solutions (Norcross et al., 2011).

In four short years the Emmanuel Movement changed the perspective of the medical community in regard to Psychotherapy and its benefits (Norcross et al., 2011). However, neurologists also became increasingly aware of the possible competition that they faced if this movement became a more nationally praised movement (Norcross et al., 2011). Thus, they largely rejected the Emmanuel Movement and suggested that it was not rooted in any scientific knowledge and was not of 'sound religious knowledge' (Norcross et al., 2011). The Emmanuel Movement is known as the main factor that led to the medical/psychiatric community announcing control over the renowned treatment method (Norcross et al., 2011).

BOSTON SCHOOL OF PSYCHOTHERAPY AND THE INAUGURATION OF CLINICAL PSYCHOLOGY

In Boston a group of psychologists and physicians were slowly working their way through conducting research based in Psychotherapy (Norcross et al., 2011). It also gave way to a prominent number of physicians and psychologists to start using Psychotherapy in a clinical context (Norcross et al., 2011). This group is known as the Boston School of Psychotherapy (Psychopathology) (Norcross et al., 2011). The research conducted within the school originates from French psychopathologists and is influenced by the spiritual and mental healers' perspective of Psychotherapy (Norcross et al., 2011). The studies of this school included research about Nervous Disorders, Hypnosis, Suggestion, dissociation, and the subconscious (Norcross et al., 2011). Although Psychotherapy is commonly tied to Freud, the Boston School of Psychotherapy were the original researchers of a lot of the concepts that are directly correlated with (Norcross et al., 2011). For example, Adolf Meyer, a prominent figure in the school, was one of the first to tie psychiatric issues with physiological features which he named 'Psychobiology' (Norcross et al., 2011).

Lightner Witmer (1867-1956), in 1896 created the first psychological clinic which already established him as a leader in the psychological science community (Norcross et al., 2011). Then, in 1907 he went on to manifest the first scholarly psychological journal known as The Psychological Clinic (Norcross et al., 2011). He systematically and scientifically used methods of diagnosis, prevention and treatment of mental disorders (Norcross et al., 2011). His research mainly consisted of working with mentally and morally retarded children (Norcross et al., 2011). However, he suggested that his methods could be used across all different types of individuals, including adults (Norcross et al., 2011). He was one of the most prominent advocates of Clinical Psychology and his works still have a high magnitude of influence on Clinical Psychology as a field today (Norcross et al., 2011).

Conclusion

Psychotherapy was originally the product of treating mentally ill patients kindly because Psychology has an apparent history of mistreating countless individuals (Norcross et al., 2011). Some exceptionally compassionate individuals attempted to correctly treat individuals with mental afflictions, and this became known as moral treatment (Norcross et al., 2011). Hypnosis was popularly practiced by Franz Anton Mesmer who became the catalyst of a worldwide movement of developing and practising the technique (Norcross et al., 2011). Mediums popularized Psychotherapy by offering counselling to clients (Norcross et al., 2011). 'Mentalists' advocated for Psychotherapy by emphasizing that the mind could cure all diseases mental and/or physiological (Norcross et al., 2011). However, Psychotherapy was not regarded as a scientific phenomenon until the Emmanuel movement, where psychologist Elwood Worcester, a Christian Minister practised the techniques within Psychotherapy to try to cure individuals with Nervous Disorders (Norcross et al., 2011). Finally, the Boston School of Psychotherapy conducted the first scientific research studies on psychotherapy and Lightner Witmer was one of the initial establishers of clinical psychology as a renowned field (Norcross et al., 2011).

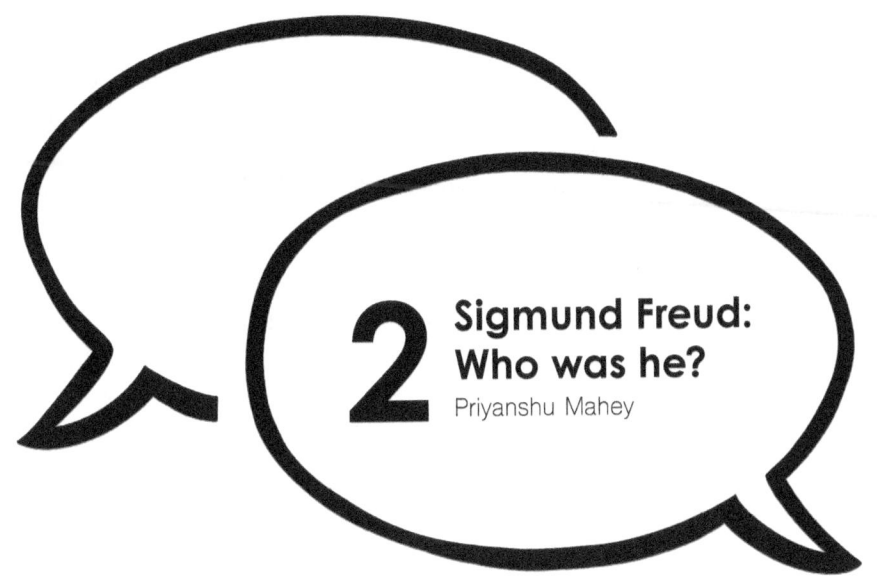

2 Sigmund Freud: Who was he?

Priyanshu Mahey

ABSTRACT

Sigmund Freud (1856 - 1939) is one of the most famous neurologists in history and is often known as the father of psychotherapy. He held many different complex theories revolving around attempting to understand the human mind. This chapter covers his history, from his birth to his marriage to his death. Freud had many interesting theories, and while not many have withstood the testament of time, have gone on to encourage other researchers and advance the field of psychology significantly. Many of his theories have led generations of scientists and helped researchers to understand the human brain better.

INTRODUCTION

Sigmund Freud, an Austrian neurologist, is well renowned as the father of Psychotherapy. Born on May 6th, 1856 in Freiberg, Moravia, in the Austrian Empire, he lived to be 83 years old. He died on September 23rd, 1939, in London England from a lethal dose of morphine from physician-assisted suicide. His theories in psychology have left an incredibly profound impact in the field and have a sustained presence in many of our modern day theories. Modern day scientific advances have proven that not all of Freud's theories were accurate and many of them were simply absurd and lacking in credibility. In this chapter, we cover Freud's background and some of his theories which gave him the title of father of psychotherapy.

Personal Life

Freud was born in 1856, and his father, Jakob, was a wool merchant. Freud's mother was actually Jakob's second wife, Amalie. He was born within the Austro-Hungarian Empire, and the area he was born in is now known as Příbor, located in the Czech Republic. His parents were Jewish, but Freud was not actively practicing in the religion (Bradford, 2016). For economic reasons, Freud's family moved from Freiberg to Leipzig and then finally settled in Vienna, where he would attend university to study medicine in 1873. Then, in 1882, Freud took a role as a clinical assistant in the Vienna General Hospital, working with psychiatrist Theodor Meynert and a professor of internal medicine, Hermann Nothnagel. He worked here until 1885 when we went to Paris for a year to study neurology as a student of Jean-Martin Charcot. Charcot, at the time, was famous for his use of hypnosis to treat hysteria. A year after his journey to Paris, Freud fell in love with Martha Bernays. He fell hopelessly in love with her and to marry her, he needed money. With academics showing little potential to bring in the required money, he became a doctor and started his private practice revolving around the nervous system and its disorders (PBS, 2002). This enabled him to marry Martha, and they eventually had six children together, one of whom would even go on to enter the same field as her father. Later on, Freud would obtain a role as a neuropathology professor at the University of Vienna in 1902, holding the position until 1938 (BBC, 2014).

In 1938, the Nazis took hold of Austria, publicly burning Freud's books, causing him to move out to London. Freud took his wife and children, fleeing for their safety. Here, Freud died a year later, on September 23rd, 1939, due to Cancer. Freud had fought against mouth cancer since 1923 and had developed a tumor that caused him excruciating pain. This pain, eventually, was too much for Sigmund Freud to bear. On September 21st, 1939, he asked his friend Max Schur to administer a fatal dose of morphine, causing him to fall into peaceful slumber until he died days later, on September 23rd, 1939 (Cherry, 2020a).

Chapter 2

EARLY WORK

When Freud had returned from Paris, he began to work alongside a physician named Josef Breuer. Sigmund Freud well-acknowledged Breuer. In 1880, he had managed to relieve hysteria in a patient using hypnosis. His work, alongside the case of Anna O., was significant in helping to form Freud's theories (Britannica, 2021). Anna O. was a woman named Bertha Pappenheim. Her hysteria was a collection of symptoms ranging from vision issues, headaches, paralysis, and hallucinations. Breuer was able to use what was considered to be hypnosis at the time to ease her symptoms. Pappenheim dubbed this the "talking cure," and Freud was incredibly fascinated by it. Down the line, Breuer and Freud would disagree on the exact specifics of how this worked, but this was an essential step in Freud's career (Cherry, 2020b). Pappenheim's hysteria would later worsen, and one of Freud's disciples had written in his book that this was not the outstanding success that many had assumed it to be (Jung 1989).

Anna O., on her own, was a fascinating figure, and the hysteria related around her and her cure has been put into question. While she was growing, she showed no signs of neurosis and was noted for having high intelligence and quick to pick up on skills. She was reported to show signs of hysteria only once her father died. Thereupon Anna O. challenged her with many different complications. Dr. Breuer then treated her in three distinct phases. The first phase was through him telling her fairy tales. The second phase was by hypnotizing her every morning. The third phase was when Anna would add descriptions to the various occurrences responsible for leading her to hysteria. Her story came into question when it was found that she was admitted to a sanatorium, remaining in a distributed start after she was supposed to be cured. Despite being the person pushing Breuer to publish research on this, Freud was also the first to criticize Breuer for ignoring obvious red flags in this experiment. In addition, Pappenheim was said to have sexual feelings towards Breuer, introducing complications in the experiment.

Freud's experiments and his theories would rely heavily on finding his own "talking cure," but this time, he found it more logical and reasonable (Launer 2005; Wayne 2016).

FREE ASSOCIATION

Freud was only able to begin grasping Breuer's story's implications on the psychological landscape once he developed a unique technique known as free association. Set during 1892-1898, free association is a therapy technique where the patient can share whatever thoughts come into their mind. Freud published a joint paper with Breuer about this in 1895 named Studien über Hysterie (Studies in Hysteria) (Jay 2021). The key idea was to have patients share whatever came into their minds, no matter if it was a random jumble of words or not. This would be linked to the patient's unconsciousness and could be used by the therapist to understand what was going on in the patient's mind. By saying whatever came to mind, they would overcome the mind's resistance. By getting past the mind's resistance, the patient can say whatever they would like without censoring themselves. This view into the patient's unconscious mind is meant to provide the therapist with a good idea into the patient's mind (Jones, n.d.).

Free association is not without limitations. Freud became insistent that there was a sexual link between the resistance faced by a patient. The patient would have sexual desires and feelings, which the logical mind would counter. He believed this to be an essential step to understand to overcome the hysteria (Evan, 2021). Unfortunately, these beliefs had very little scientific evidence to back them up. In the modern-day, free association is less frequently used and has changed dramatically since the way it was initially used. One of the most significant shortcomings of the theory is that the patient cannot describe what they do not yet comprehend, whether consciously or unconsciously. Another shortcoming is that Freud wanted to lower the number of suggestions presented to the patient. This, however, would be gravely impacted once the therapist comes into question. The interactions with the therapist would alter the conversations and change the overall way the patient interacted. The things a therapist does will inevitably affect the patient's behavior (Marmor, 1970).

Another big issue with the free association was that it was heavily based on the desires of Freud. Therefore, it was best suited to help Freud and, in nature, was not as generalizable as therapists would hope it to be.

Nowadays, this technique holds a lot less weight, being used significantly less in the original manner described by Freud. Instead, the modern-day free association makes more use of the interactions between the patient and the therapist, with the therapist guiding and engaging with the patient more (Schachter 2018).

Development of Psychoanalysis

After coming up with his original theories, Freud began to leave the term hypnosis, and by 1896, he was using the term psychoanalysis. The first occurrence of this term was in his article called "L'hérédité et l'étiologie des névroses" which translates to "Heredity and etiology of neuroses" first published in French in 1896 (Freud, 1896). The same year, he published his seduction theory. The seduction theory states that hysteria was a side effect of sexual

Assaults while they were younger. Freud would later leave this theory for reasons not entirely known. While some suggest that this was because it led to disapproval from his peers, others believe there were more complex reasons for his beliefs on his theories. The seduction theory has received mass criticism, and one particular reason was that its results were indisputable. This was because Freud had not described any methods to reproduce them. This is probably the biggest issue with the seduction theory. It becomes impossible to prove it wrong without knowing how to derive conclusions (Israëls, & Schatzman, 1993).

To develop onwards, Freud started to analyze himself to understand his theories better. He was eventually able to create the idea that much of the neurosis seen in people was due to early trauma, which ultimately led to him developing ideas about sexuality and repression, which would become prevalent in much of his later work (Burton, 2015). On October 23rd, 1896, Freud's father, Jacob Freud, died at the age of 81, leading to Freud experiencing complex and perplexing emotions. These emotions led Freud to seek answers as to why he felt this way, and it led to him deeply analyzing himself. This was when he realized he had replaced affection towards his father from his mother and targeted it towards himself. He began imagining that his father was abusive, and through all this, he was able to come up with the Oedipus Complex. The Oedipus Complex states that a boy would want to marry their mother, and their father would be seen as a rival for her love (PBS, 2002). All this work would become significant pieces of his legacy.

Freud got to experience his fair share of exciting experiments. Few specific cases are known for contributing to his research career and are reasonably attractive. One of them is the case of "Little Hans." Little Hans was a five-year-old boy with an odd phobia of horses. Freud learned that little Hans was interested in penises, and to get him to stop playing with his own, his mother would threaten to cut his penis off if he did not stop. Freud linked this to the fear of horses because horses have large penises. The horse phobia then shifted and became a fear of only horses with black harnesses. Here, Han's father related this to his mustache, and subsequently, Freud was able to conclude. Freud proposed that this was the Oedipus complex, and Hans was simply in a contest with his father. Hans himself wanted a large penis and wanted to take his father's position by marrying his mother. This conclusion enabled Hans to calm down and get control of his fear and become closer with his father (McLeod, 2008).

Another interesting case that Freud explored was "Rat Man." This name was given due to the patient's nightmares revolving around rats. The exact identity of this person is still up in the air for debate, but as of now, most people tend to believe this person is Ernst Lanzer. Ernest Lanzer was often terrified that something would happen to him, his father, and his female friend that would go on to become his wife. The exciting thing about Lanzer's case was that after going to therapy for six months, he had been cured, and this happened to be the first time that psychoanalysis had resulted in a cure. Sigmund Freud described his case as obsessional neurosis at the time, and nowadays, it is called obsessive-compulsive disorder. After doing some analysis on the man, Freud was able to figure that the man's thoughts originated from the military, where he had learned about a form of torture from rats. Freud was further able to know that the man was sexually intimate while younger, and Freud believed that neurosis had already taken hold of him during his childhood. Lanzer's desire to see a woman naked while he was younger was an obsessive wish he held, and in response, he developed an obsessive fear. Again, Freud linked this to the penis because rats carry infections, and pensies can have syphilis (Thapaliya, 2017).

One more case that helped Freud develop his theories was Dora, an eighteen-year-old girl suffering from hysteria. Her father brought her to Freud, hoping to get her cured. Freud, of course, theorized her hysteria symptoms to be related to psychological trauma related to her sexual life. It turned out that when she was a child, she was fond of her father, who took care of her while she was sick. She happened to become close friends with a married couple, Herr and Frau K. Herr K. and Dora developed a close bond, eventually to the point where Herr proposed to Dora. Dora went home and told her father, but as it turned out, Herr K. denied the proposal, and her father assumed that Dora had imagined this event. Herr K. had also kissed her by surprise, which left Dora disgusted, and when she prompted her father to stop their relationship with the married couple, his father brought up that they were indebted to Frau K. as she helped take care of her while she was sick. Dora suspected there to be a love affair between the two. Again, Freud focused explicitly on the dreams Dora had.

One of the first dreams he focused on was the dream where Dora's house is on fire, and she attempts to wake her parents. Her mother looks for her jewelry box while Dora, her sisters, and her father try to escape. The father mentions that he will not attempt to save the jewel case and her father essentially saves the family from the fire. Freud noted that jewel case was slang for vagina, which was potentially representative of her concern that Herr K. might do something to her and her father would not stop it. Freud saw this as absurd since it was a sexual moment to have felt attracted. To him, it seemed like Dora was repressing her attraction towards Herr K. The second dream Freud observed was the dream where Dora walks into a strange town, arriving at her current home, and when she entered her room, she saw a letter. This letter was from her mother, and it stated that her father had died. She attempted to get to the funeral but could not find the train station. Suddenly, while looking for the station, she was teleported home to find that her mother had already left to go to the funeral. Freud argued this was a revenge fantasy against her dad (Billig, 1997).

THE INTERPRETATION OF DREAMS

In 1899, Freud published his crowning achievement where he put forth findings from his clinical practice. This is considered by many to be his masterpiece as it contained many of his theories that would leave a significant impact on the field. The interpretation of Dreams describes dreams as a look into the world of a person's unconsciousness. They were a result of the unconscious mind trying to express itself and get past the body's resistance and the conscious mind. Here, he also described the Oedipus Rex complex as a topological model of the mind. This book was highly influential in discussing the state of the reason (Burton, 2015). Dreams, to Freud, showed desire and were a fulfillment of people's wishes. With this book, he hoped to help others to decipher dreams.

According to Freud, dreams can be split into four primary activities. First comes condensation, which essentially combines different aspects of psychic life to create unique, meaningful symbols. The second activity of dreaming is displacement which is essentially decentring of thoughts. This essentially means that the person's wish may not even be at the center of their dreams. This means that they might be focusing on a non-important aspect of their thoughts while what they want is simply a background element. The third activity is a representation which is how dreams become visuals. Lastly, there is a second revision that makes dreams more coherent and more story-Esque (Evan, 2021).

SIGMUND FREUD'S LEGACY

While what was discussed so far is by no means his entire collection of research, this is a pretty good overview of some of his significant contributions to psychoanalysis and the field of psychiatry as a whole. However, that is not to say his theories were all correct and many of them are highly disputed. Freud has faced his fair share of criticism, and his role as the father of psychoanalysis has also been put into question. Freud's modern-day reputation varies. Many feminists and psychologists have noted Freud's views on women. He tends to have a sexist idea of them, seeing them as less than men in many cases. His theories were often sexist, and his misogyny was pretty prevalent in his approach towards hysteria. Many of Freud's theories were based on his understanding of his mentality, and it lacked a feminine perspective. This inability to include the view of a woman left a misogynistic attitude in many of his theories (Nolan, & O'Mahony, 1987; Schafer, 1974).

Freud's theories were often sexual and made many at the time uncomfortable. His peers would often disagree with his theories in part due to their highly sexual nature. While it left many of his peers to feel uncomfortable, his research into sexuality opened a pathway for other researchers to explore and study the field deeper. His outlandish and sometimes absurd theories opened conversations that are still relevant today, and this conversation has helped to form a scientific understanding of sexuality (Ellism 1939).

Perhaps Freud's most significant and most positive impact was in clinical psychology. Freud had thoroughly changed the field with his version of the "talking cure," and psychoanalysis went on to inspire many future researchers and therapists. This method has become synonymous with psychology, and while fewer therapists use it now, it still holds a significant impact. Attempting to understand their patients and discuss their complex issues to get to the underlying truth beneath them is still relevant today and is critical for the entire field of medicine in general (Murphy. 1956; Ruitenbeek, 1967).

CONCLUSION

Without Freud's enormous progress, psychoanalysis would not have been the way it is now. He's grown a lot, and by studying his history, it becomes apparent how much of a prominent figure he was in the psychiatric community. Not only did he leave his impact in psychology but other fields of medicine and philosophy again. The strength of his research's impact is undeniably immense.

What is Psychotherapy?
Benjamin Turner

Abstract
Psychotherapy is a wide-ranging, complicated subject from the initial principles discussed in Chapter 2 through the works of Dr. Sigmund Freud and the subsequent century of research and practice that he inspired. This section discusses psychotherapy to build a basic understanding of the subject, including an overarching definition of psychotherapy, basic principles, elements, and a method of reviewing the levels of psychotherapy.

Introduction
Therapy evokes images of a stuffy office where the patient lays on the couch and talks about their problems while a therapist sits nearby with their legs crossed and takes notes. However, the reality is far more complicated than a patient simply complaining about their life. The practitioner must earn the patient's trust, make them feel comfortable, create an environment where they don't feel judged, so they have the freedom to unburden themselves (Bruch, 1974).

Psychotherapy is unique in that it is more collaborative. The therapist does not work on a patient. Instead, they work in conjunction with a patient; their role is to facilitate, not operate. The methods and approaches are numerous and covered in greater detail in Chapter 4, but they are tailored to the individual as a bespoke suit. This chapter will introduce a general definition of psychotherapy and some essential elements and principles.

DEFINING PSYCHOTHERAPY

In its most basic elements, psychotherapy is communication between a therapist and a patient that attempts to help the patient feel relief from emotional stress, seek solutions to problems in their lives, and modify their thoughts and actions in a constructive manner (American Psychological Association, 2017). Psychotherapy is often colloquially referred to as talk therapy because treatment consists of a conversation with a trusted advisor. Most people undergo psychotherapy every day just in talking with friends and loved ones. Still, there are some essential differences between getting something off your chest with a friend and receiving professional psychotherapy in a clinical sense.

Talking about problems with friends is an essential part of mental health, but it has its limits. There are two important distinctions to psychotherapy with a licensed therapist. The first is that patients speak to a trained and licensed professional with experience identifying stressors and treating patients. The second is that social relationships have a give and take nature; the relationship between friends balances the usefulness of that friendship to both parties. In other words, friends are not likely to hang around for long if one party is always in need and the other is constantly tending to them. In professional psychotherapy, the focus is entirely on the patient's needs, and the treatment program hinges entirely on those needs (Bateman, Brown, & Pedder, 2000). It is a standard practice for the doctor and patient to agree to a treatment contract, either verbal or written, that outlines their goals, establishes a regular schedule for time, place, and duration, and defines procedures for treatment (American Psychological Association, 2017).

Trust plays a central role in a therapeutic relationship. For therapy to be successful, it is essential to develop a strong bond (Bruch, 1974). This can be challenging because the prospect of bringing their problems to a stranger will naturally stir up anxiety in the patient. They may be concerned that the therapist will feel their time is being wasted or that they are being judged. These are healthy and normal concerns. Once these standard barriers to trust are overcome, the therapy can begin (Bateman et al., 2000).

Mainly because psychotherapy is the treatment of the mind, not an individual physical ailment, it becomes clear that without trust, the effectiveness of therapy diminishes. Therefore, a therapist must be skilled at creating a constructive relationship with their patient. This concept is known as building an alliance; there must be the mutual impression that both parties are on the same side working toward a common goal (Bateman et al., 2000). This helps to paint a clear picture of what is seen in research: that patients who have a good relationship with their therapist have a dramatically reduced dropout rate (Tsai, Yoo, Hardebeck, Loudon, & Kohlenberg, 2019). So naturally, patients who maintain the entire course of prescribed treatment are much more likely to achieve the goals outlined in their treatment contract.

Trust in psychotherapy must also include an element of confidentiality. This is true of all medical treatments identified as early as Hippocrates. A unique relationship exists in psychotherapy because, unlike other disciplines where a doctor is actively working on a patient, psychodynamic therapy requires the doctor and patient to work together. They must engage one another in a shared space and attempt to mitigate and resolve the effects of past trauma (Bateman et al., 2000).

Effective therapeutic listening is an active process of alert and engaged participation in the struggles of what ails the patient (Bruch, 1978). The element of alliance between doctor and patient is not an easy one because to be successful, a therapist must be able to strategically challenge and confront their patient (Bateman et al., 2000). The relationship will be strained, but a skilled practitioner can identify how far and how hard to push and effectively balance the patient's trust with the progress of their treatment.

The concept of transference must also be discussed. This occurs when a patient projects their unconscious feelings or wishes, previously directed to a vital individual such as their mother, onto the therapist (APA, n.d. b). In psychotherapy, this concept is a valuable tool, thought to help evoke suppressed experiences or offer insight into how individuals process events. It can be thought of as the unconscious repetition of behavior and projection onto a new subject.

There exists an additional element to transference as well. The therapist is only human to project their unconscious wishes or feelings onto their patient; this is known as countertransference (APA, n.d. a). Once thought of as something to be avoided, the more contemporary approach is to recognize that countertransference is, to some degree, inevitable. However, it can be viewed positively to understand the effect the patient has on other people. The important part is to be aware of the phenomenon and find some use for it if at all possible.

These are some of the critical elements of clinical psychotherapy. First, it is a structured conversation between a licensed professional and a patient seeking help for any number of non-physiological issues. Second, the therapist works with the patient, not on them, to identify and address their complaints; this requires the two parties to form an alliance where they feel they are both working toward a common goal. As discussed in chapter 2, these methods were pioneered by Dr. Sigmund Freud, who subsequently inspired an entire field of research. (Bateman et al., 2000).

Principles of Psychotherapy

There are different approaches and forms of psychotherapy. However, there are two primary streams: psychodynamic psychotherapy as founded by Sigmund Freud and behavioral psychotherapy first explored by Ivan Pavlov. These two streams have evolved and found themselves occasionally working in conjunction and at odds with one another. The approach in the psychodynamic stream is empathetic, one of helping the patient recognize their triggers and stresses internally and attempting to fix them consciously; more simply, this approach can be thought of as fulfilling the ancient Delphic instruction to 'Know Thyself' (Bateman et al., 2000).

In the behavioral approach, the focus is more on studying and influencing the patient from the outside. For example, to observe behavior analytically and introduce stimuli to reward or punish those behaviours based on some agreed-to goal. This approach has joined with cognitive science and is referred to as cognitive-behavioural psychotherapy (Bateman et al., 2000). In this book, the psychotherapy we discuss is through the lens of the research inspired by Dr. Sigmund Freud, as discussed in Chapter 2. Cognitive-Behavioural psychotherapy will be discussed further in Chapter 5.

Psychodynamic psychotherapy has various approaches and schools of thought within it, but they share a few key concepts. These include:
When a patient experiences conflict stemming from unacceptable aspects of themselves or their relationships, they may seek help in addressing that conflict (Bateman et al., 2000).

Freud was the first to suggest that a patient may be so disturbed by an aspect of themself that their conscious mind rejects it, and the aspect is expressed subconsciously. The result of this subconscious expression is that the patient may experience anxiety or psychic pain but is unaware of the cause. People all employ defence mechanisms to deny, disown or suppress what is unacceptable to consciousness. These mechanisms can sometimes be unhelpful or even harmful to the individual (Bateman et al., 2000).

Wishes, feelings, or memories that are unacceptable to the individual may surface in response to basic motivational drives, creating conflict. These factors often form during the different developmental phases of a person's life. The concept of basic human drives is controversial in its details, with different psychodynamic schools subscribing to different theories on how to categorize them and which are most problematic, but the central idea that motivational drives are a source of conflict is common to all schools (Bateman et al., 2000).

The most debated concept between psychodynamic schools of thought stems from models of the mind. Freud initially theorized about conscious and unconscious, later adding ego, super-ego, and id. Berne theorized about each of us' parent, adult, and child levels. What is agreed about is the concept of psychic levels and that conflict can occur between them (Bateman et al., 2000).

Finally, there are aspects of the therapeutic relationship that also affect the mental health of the patient, including alliance, transference, and countertransference (Bateman et al., 2000).

The goal of psychotherapy, as first explored by Freud, is to explore a patient's conflict through these lenses. First, to understand their psychic pain, the therapist attempts to identify the cause of that pain and then explore how the cause functions through this lens of the models of the mind, motivational drives, developmental phases, and how it is being affected the therapy itself.

LEVELS OF PSYCHOTHERAPY

There are different ways of expressing the progression of psychotherapy, but for the sake of simplicity, we will discuss the levels of psychotherapy in the terms laid out by Cawley (1977). He put forward a four level system for recognizing psychotherapy, with steps one through three focusing on psychodynamic theory and step four aimed toward behavioural psychotherapy; this chapter will focus on steps one through three. The general logic of this system focuses on the professional training and experience of the therapist and the resulting therapy they are qualified to offer, as well as the intensity of the therapy. What Cawley outlined was the higher the level, the deeper the exploration.

Psychotherapy I is what we find with the average doctor or medical professional. It is the process of recognizing and empathizing with the complaints of the patient. Initiating the communication process and identifying whether or not a referral to a mental health specialist is necessary. If a referral is made, level one also involves helping the patient to feel comfortable with the referral. As mentioned earlier in this chapter, there are all sorts of anxieties that crop up when someone is sent to a therapist. The medical professional at level one should recognize this anxiety and attempt to dispel them to the extent possible (Cawley, 1977). If a patient does not need to progress beyond psychotherapy I, this is sometimes referred to as outer level psychotherapy (Bateman et al., 2000). What the patient needs is a sympathetic, trusted listener, to help them unburden themselves from their feelings; for many people, this is all they require to live a healthy, fulfilled life.

Psychotherapy II happens with the psychiatrist, social worker, or other mental health professional. It encompasses some features of level one, but the professional must be able to understand and communicate with patients suffering from any sort or degree of psychological conflict (Cawley, 1977). This requires recognition that the present attitude and state of the patient are characterized by previous experiences and their way of thinking about the world and themself; these factors are often outside of the patient's awareness and control. At this level while transference or the identification of cognitive assumptions may be recognized, they are not generally commented on. Instead, they are used to understand the patient better, or perhaps employed to strengthen the therapeutic alliance.

Psychotherapy III is what many people colloquially think of as psychotherapy, particularly in the formal sense. This level includes the characteristics from levels one and two around acceptance, understanding, and respect, but there is a greater focus on getting the patient to actively recognize the causes of their conflict. Level three goes deeper still in employing specialized techniques to help the patient understand the factors of their problems, the aim being for the patient to achieve a level of agency so that they can take responsibility for themselves and their relationships (Cawley, 1977).

Conclusion

Psychotherapy is a complicated and wide-ranging subject, but it's most basic elements are fairly universal. It is simply a structured conversation wherein collaboration with a trained professional, the patient works to identify the causes of their psychological conflicts. To face this challenge, the therapist must quickly build a constructive relationship with the patient, recognize the type and degree of conflict they are facing, and identify the level and type of psychotherapy that is best able to help them. They must juggle their alliance with the need to ask probing questions and confront the patient where it will ultimately be beneficial, while maintaining an awareness of both transference and countertransference.

There are various types of psychotherapy that fall under this umbrella that will be discussed in detail in Chapter 4. They are numerous and range from individual psychotherapy to couples and group therapy, and there is a distinction between psychoanalysis and psychodynamic practices. The methods and approaches are legion, but they all begin with the foundation laid out in this chapter; trust, acceptance, empathy, understanding, and collaboration are the bedrock of all psychotherapy.

4 What are the types of Psychotherapy?
Neha Saroya

ABSTRACT

The purpose of this chapter is to provide a brief overview of the five categories of psychotherapy as outlined by the American Psychological Association: psychoanalysis and psychodynamic therapies, behaviour therapy, cognitive therapy, humanistic therapy, and holistic therapy. Each type of therapy is defined by specific principles that guide the techniques used by therapists that practice them. Many of these categories involve a particular set of beliefs about the world, ourselves, and others and how our emotions may impact our actions. A large majority of therapists will choose to use holistic therapy, which combines elements of different types of treatment to best suit the needs of their clients.

INTRODUCTION

Although there are over 50 types of therapeutic approaches that mental health professionals can use when treating mental illness or addressing mental health concerns, there are five broad psychotherapy categories (Rauch, 2016). In addition, many organizations categorize forms of psychotherapy in different ways, and the categories discussed below are outlined by the American Psychological Association ("Different approaches to psychotherapy," 2019). The form of therapy that a therapist chooses to use should be carefully selected upon analyzing their client's concerns and needs.

Psychotherapy Categories
Psychoanalysis and psychodynamic therapies

Psychoanalysis therapy techniques stem from the work of Sigmund Freud, although it has evolved drastically since its original conception ("Different approaches to psychotherapy," 2019). Psychoanalysis refers to what you would typically think of when you hear the word 'therapy' where a client undergoes long-term therapy with regular sessions ("Different approaches to psychotherapy," 2019). The focus of the sessions is talking through one's feelings and emotions in an attempt to reveal unconscious feelings to the forefront of the mind so the client can start to work through them ("Psychoanalytic Therapy," 2021 - A). Often, the emotions result from things that occurred in the clients' childhood that the client has since repressed as a form of self-protection ("Psychoanalytic Therapy," 2021 - B). The therapist will then help the client navigate how these feelings affect the way they act and think and how it impacts their relationships with others. Those with common conditions such as generalized anxiety and depression or even specific emotional concerns like relationship difficulties are well suited for this type of therapy ("Psychoanalytic Therapy," 2021 - B). The therapist's role in this setting is to guide the client to understand their inner self by helping them interpret their feelings and actions. Once this is achieved, the therapist can work with their client to develop healthy coping mechanisms so that they can effectively deal with their current and potential future situations ("Psychoanalytic Therapy," 2021 - B).

Psychodynamic therapy draws on similar techniques as psychoanalytic therapy, focusing on introspection and assessing how the mind works (Shedler, 2010). However, several key differences make them utterly different therapy techniques. For starters, psychodynamic therapy tends to be much briefer than psychoanalytic therapy, which is often long-term and over the years. Psychodynamic therapy is typically once a week, and it is not unusual for it to only last for up to 15 sessions. However, the number of sessions can be left open-ended (Shedler, 2010). Additionally, there is less of a focus on the work of Sigmund Freud, and his ideas are rarely implemented in psychodynamic therapy today. Finally, there are seven psychodynamic therapy techniques that distinguish it from other forms of treatment (Shedler, 2010).

- "Focus on affect and expression of emotion": this refers to an exploration of all of one's emotions and the therapist's role in helping to verbalize these feelings.
- "Exploration on attempts to avoid distressing thoughts or feelings": avoidant behaviour is actively focused on in therapy sessions to try to get to the root cause of avoidance and the emotions attached to it.
- "Identification of recurring themes and patterns": the therapist helps the patient to identify patterns in their thoughts that they may not be aware of in attempts to recognize and break harmful patterns of behaviour.
- "Discussion of past experiences": the focus is mainly on one's childhood so that they can see how the past affects their present thoughts and behaviours.
- "Focus on interpersonal relations": often, problems can arise when interpersonal relations, such as how we interact with ourselves and others, do not adequately meet one's personal needs — taking a closer look at these relationships can help identify gaps to be worked on.
- "Focus on the therapy relationship": the relationship between a therapist and their patient is an important one, and understanding how the patient's interpersonal relations play into their relationship with their therapist provides a chance to work through them in vivo.
- "Exploration of fantasy life": a key trait of psychodynamic therapy is to allow the patient to speak about whatever they are thinking about in that moment, which often leads to a discussion about dreams and fantasies. Insight into these topics can help the therapist understand how the patient feels about their own lives and what may be interfering between the patient and their ideal life.

Behaviour Therapy

Behaviour therapy emphasizes how one acts rather than how one thinks

("Different approaches to psychotherapy," 2019). The goal of behaviour therapy is often to change the way one works, and there are four key characteristics of this type of therapy (Eelen, 2018).

- "The assumption that abnormal or problem behaviour is learned": This is the first and most commonly known characteristic. Behaviour therapists believe that the client's behavior to change was learned at some point, usually during childhood, in the same manner, that 'normal' behaviours are learned.
- "Behaviour therapy is an application of findings from experimental psychology": This principle allows the field of behaviour therapy to be ever-changing as experimental studies develop new findings. Rather than being stagnant, behaviour therapy was created with the opportunity to evolve in mind constantly.
- "Direct focus on the behaviour": When behaviour therapy was created, it focused mainly on the behaviour at hand that the client wishes to modify or change. It has now expanded, and many therapists also attempt to delve into the root causes of said behaviours.
- "Methodology": Behaviour therapy was created with a specific experimental methodology in mind that could be used to change abnormal behaviours. This starts with the therapist making observations about the behaviour, then developing a hypothesis of why they believe that behaviour exists, followed by selecting the best intervention to modify the behaviour.

As behaviour therapy has grown in the past 25 years or so, many specific techniques have come about that therapists use depending on client needs. For example, cognitive-behavioral therapy (CBT) is a prevalent technique that combines both behaviour therapy and cognitive therapy, which is discussed in the next paragraph (Gotter, 2018). CBT is used when someone wants to change both how they think and how they act; thus, treatment focuses on problem-solving and understanding how the client's emotions impact their actions (Gotter, 2018). This can also be used with children, and it is called cognitive behavioural play therapy. In this therapy method, the therapist will watch how a child plays and interacts with their environment to see how they express their emotions and what makes them uncomfortable ("Cognitive Behavioural Play Therapy," n.d.). This is especially effective since most children cannot verbally express their emotions or experiences, mainly if they have experienced something traumatic or something they perceive to be shameful. Cognitive behavioural play therapy can help the child learn how to communicate better and provide a safe space for them to release emotions ("Cognitive Behavioural Play Therapy," n.d.). Another technique that is often used to treat phobias is called system desensitization (Gotter, 2018). This technique is based on the principle of classical conditioning, which is further explored in the next chapter. The therapist provides a safe space for the client to face their phobias head-on and will teach them breathing or relaxation techniques that can aid them when exposed to the phobia. Then, these techniques are practiced as the therapist exposes the phobia to the client in small doses that increase over time, with hopes that eventually the client will be able to deal with the phobia in the real world (Gotter, 2018).

Cognitive Therapy

In cognitive therapy, contrary to behaviour therapy, the main focus of the session is the clients' thoughts rather than their actions. Therapists who practice this form of therapy believe that thoughts are more indicative of mental state than actions are ("Different approaches to psychotherapy," 2019). This means that even if someone has a seemingly stable life, with support systems in place and well-developed relationships if their thoughts are dysfunctional, it could lead to problems. Thus, it is necessary to get to the root cause of these dysfunctional thoughts in order to try and alter this line of thinking. This can oftentimes be very difficult as these thoughts could be living in the clients' minds for years or even decades. Thus, it is critical for the therapist to help guide the client at their own pace to help them understand how to turn dysfunctional or negative thoughts into something more positive ("Different approaches to psychotherapy," 2019). Compared to other forms of therapy, such as psychoanalysis, cognitive therapy can be relatively brief ("Definition of Cognitive therapy," n.d.). It also has a focus on how the client thinks, communicates, and feels in the present, rather than dwelling on the past ("Definition of Cognitive therapy," n.d.).

HUMANISTIC THERAPY

Humanistic therapy places a focus on the client and their concerns, while the therapist's role is to actively listen to these concerns and allow the client to guide the session (Raypole, 2019). This type of therapy places emphasis on the client's day-to-day life and is rooted in the belief that intrinsic fulfillment can be achieved by becoming one's "true self" (Raypole, 2019). Humanistic therapists believe that at the core, everyone is a 'good person and can choose to make decisions that will benefit them, but this can be difficult to accomplish for those who do not believe in themselves or their capabilities ("Different approaches to psychotherapy," 2019). This stems from the principle that the light in which one sees the world around them has a significant impact on their actions. As such, in therapy sessions, the goal is often for the client to understand the world around them and their place within it to try and fully accept themselves for who they are (Raypole, 2019). If the client sees themselves in a negative light, it will often affect their worldview, and by working to see themselves in a positive manner, they are effectively changing this view. The therapist will aid the client in personal growth in hopes that it will positively impact their behaviours and thoughts. Two critical aspects of achieving this personal growth are respect and concern for others ("Different approaches to psychotherapy," 2019).

There are three main types of humanistic therapy, and all of them share a common goal of helping the client to achieve their maximum potential. The most commonly used of the three is client-centered therapy, in which the therapist does not have an authoritative role (Raypole, 2019). Rather, the therapist helps facilitate the client's growth by expressing active listening skills, keen interest, and concern. It is essential to form a strong client-therapist relationship for this type of therapy to achieve maximum effectiveness. The therapist needs to trust that the client can be in complete control of their own thoughts, and the client needs to trust the therapist enough to open up to them in order to guide sessions. This bond will allow the client to talk about their experiences without any fear of judgement or disapproval because they know their therapist will accept them unconditionally ("Different approaches to psychotherapy," 2019).

The second type of humanistic therapy is Gestalt therapy, which focuses on "organismic holism." This refers to being completely aware of the present and everything going on in one's life ("Different approaches to psychotherapy," 2019). This is based on the theory that unresolved conflict, conscious or not, can lead to negative emotions, so uncovering these thoughts and accepting responsibility for your part in them can do wonders for one's mental state (Raypole, 2019). There are several techniques that a Gestalt therapist can use to help their client through these conflicts, and they largely involve recreating scenarios in a safe space so the client can discuss how they make them feel. This can be done using role-playing, reenacting scenarios, or via exaggeration (Raypole, 2019). The last type of humanistic therapy is called existential therapy, and it focuses on the true meaning of everything that happens in one's life (Raypole, 2019). Under the guidance of a therapist, the client can explore why they do certain things or why things happen to them so that they can gain a better understanding of themselves. The therapist will also help guide the client to see how their conscious and unconscious thoughts affect their actions and mental state (Raypole, 2019).

HOLISTIC THERAPY

Every client's needs are unique, and a therapist must be able to adapt to situations in which a client may need more than one therapy technique to help them reach their goals. For this reason, the majority of therapists adopt holistic therapy, which is when they combine different therapy techniques from different psychotherapy categories to tailor the therapy to each individual client ("Different approaches to psychotherapy," 2019). However, there are many therapists who will choose to specialize in specific therapy techniques that interest them and will primarily work with a client base looking to utilize that form of therapy.

CONCLUSION

The differing views of the human psyche that academics hold have resulted in a wide array of psychotherapy categories. These categories are very broad and do not serve as a template for therapists to use in their sessions but rather as a roadmap to help them assist their clients with their emotional needs. In order to effectively treat their clients, it is also imperative for therapists to be able to determine what type of therapy techniques would work best with every client, given their individual situation and past experiences. This can be done by communicating with the client to determine what specific goals they would like to achieve with therapy and what emotional needs must be fulfilled. Moreover, it is also important for a therapist to know their own limitations and not attempt forms of therapy with clients that they are not trained in or licensed to do. At the end of the day, therapists must always keep the best interests of their clients in mind.

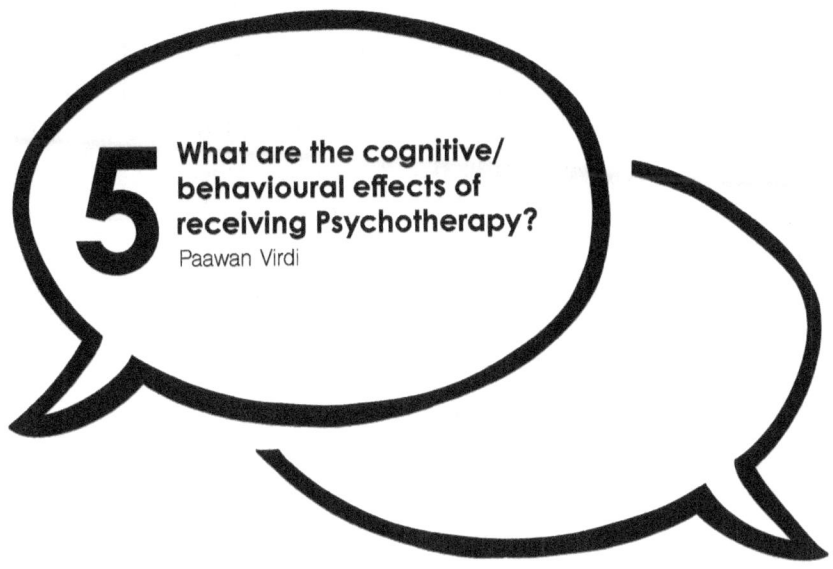

5. What are the cognitive/behavioural effects of receiving Psychotherapy?

Paawan Virdi

Abstract

The purpose of this chapter is to explore the effects that the various forms of psychotherapy discussed in the previous chapter have on the individual receiving them. In this chapter, an emphasis is played on the cognitive and behavioural effects of these treatments. The chapter begins with exploring some general effects of psychotherapy, such as its influence on neural networks and the symptoms of certain psychological disorders. The next chapter explores types of psychodynamic therapies and discusses the method known as interpretations and its role in psychotherapy. Behavioral therapy is explored as well, with an explanation of classical conditioning and how it has been incorporated in the treatment of certain conditions. The effects of Cognitive therapy and Cognitive-Behavioral therapy are also discussed in brief. The chapter then concludes with a discussion of humanistic and holistic therapies and how they guide change in the patient.

Introduction

Meta analyses have been commonly used in the field of psychology to achieve an understanding of the efficacy and consequences of certain types of therapy and medication. Such research has provided insight into how psychotherapy impacts the way one thinks, and how it influences cognition and behaviour. Cognition encompasses the mental procedures that are involved in the perception and processing of information (Dumper et al., 2019). It refers to the process of thinking; organizing the massive amount of information that resides in and flows through our mind. The term "behaviour" is related to one's actions and responses to stimuli; it is what we do (American Psychological Association, n.d.). This chapter will focus on discussing how the different types of psychotherapy discussed in previous chapters influence the cognitive and behavioural processes of the individual receiving them.

General Effects of Psychotherapy

We have learned a lot about how the brain works and its role in memory. By getting an understanding of the processes that underlie our thoughts, we have been able to address psychological issues through physiological understanding. Nobel laureate Eric Kandel described five principles that relate our physiological knowledge to practices of psychotherapy. The five principles go as follows. All mental processes are facilitated through neurons and their networks. The connections of neurons are determined by genes and their products, and thus they underlie how they function. Our experiences influence how these genes are expressed, having influences on how we develop. The process of learning alters gene expression and neural connections, influencing changes in behaviour. The last of these principles is perhaps of the most significant: psychotherapy induces changes in gene expression and neural networks that lead to long-term changes in behavior (Kandel, 1998). These principles suggest that the fundamental processes of our brain are influenced by experiences and growth, and can also be affected by psychotherapy. Many neurological disorders such as schizophrenia and depression have been tied to problems in neural networks (Goodkind et al., 2015). The problematic experiences one has that cause large amounts of stress that is tied to fundamental functions in the brain being destabilized (Malhotra & Sahoo, 2017).

The positive environment that is presented in psychotherapy favours the development of cognitive and behavioural abilities. Dissociative disorders that can arise following trauma often involve the disconnection of neural networks as a result of high stress levels. In psychotherapy, there is integration of neural networks that can enhance the cognitive and emotional capacities of the patients that are negatively influenced from such trauma (Malhotra & Sahoo, 2017). Meta analyses have functioned as testaments to the efficacy of psychotherapy, which is discussed in future chapters. However, negative outcomes of psychotherapy have also been documented. While it is difficult to examine long term changes in personality traits as a result of psychotherapy, one study examined changes in certain traits in markers as a result of therapy experiences. The study found significant increases in things like stress and neuroticism and decreases in traits like self esteem as a result of therapy experiences (Chow et al., 2017). Ultimately, the different types of psychotherapy have different focuses and approaches, yielding differing outcomes and impacts on patients. It is also important to note that some types of therapy can fall under multiple classifications of therapy, and thus subtypes of psychotherapy are not exclusive to the section in which they are found.

PSYCHODYNAMIC THERAPIES
Research suggested that psychodynamic therapies have been effective in addressing many mental health issues and physical symptoms that are related to stress. Psychodynamic therapies place emphasis on having the individual reflect and reexamining their life to find problematic patterns that contribute to the emotional distress that they feel.

One example of psychodynamic therapy is supportive-expressive (SE) psychotherapy (Center for Substance Abuse Treatment, 1999b). SE psychotherapy has been used to address a variety of disorders such as depression, anxiety, and drug-related disorders (Connolly et al., 1998). It involves using supportive techniques to strengthen the relationship between therapist and patient. This is known as the therapeutic alliance and is believed to be a very important part of psychotherapy with significant influence on the outcomes of the treatment (Ardito & Rabellino, 2011). However, there are different models of SE psychotherapy which place different priorities on what to address in the treatment process (Connolly et al., 1998).

One of the main therapeutic techniques employed in SE psychotherapy is the use of interpretations, these are statements that go beyond what the patient is aware of and aim to provide the patient with new outlooks on certain things. An example of an interpretation could be stating the patient may have developed a certain characteristic or behaviour because of a previous event or relationship in their life. Interpretations aim to provide a new perspective in looking at the problems in one's life. By listening to the interpretations of the therapist and understanding their unconscious motivations, the patient learns how they may be able to address certain problems by augmenting their behaviour (Neuman, 2015). By revealing what is just outside of the consciousness of the patient, the psychotherapist facilitates enhancement of the neural networks inside their brain. By bringing the unconscious into the awareness of the patient, nodes in their neural networks are activated that can have cascading, network-wide effects which are associated with cognitive and behavioural effects. For instance, dissociative disorders that arise from stress were discussed earlier in this chapter. The process of interpretation can be vital in addressing stress, anxiety, and maladaptive behaviours (Malhotra & Sahoo, 2017). and However, it is more subtle than just that. How the therapist provides interpretations is very important. Errors in how interpretations are performed can be frustrating for the patient and could lead to negatively affecting the therapeutic alliance. For example, interpretations being performed too frequently has shown to lead to hostility from the patient (Schut et al., 2005). However, the effectiveness of interpretations may also increase with the strength of the therapeutic alliance, which is a focus of SE psychotherapy (Gazzola & Stalikas, 2004).

Another example of a type of psychodynamic therapy is brief adaptive psychotherapy (BAP). This is a form of therapy that is often used to address personality disorders. The aim is to have the patient develop cognitive understanding of maladaptive patterns: how they came to be and how they interfere with their goals (Pollack et al., 1991). By developing such an understanding of what is internally obstructing them from their goals, the therapist and patient can work on developing adaptive behaviours that better allow them to pursue their desires. These examples of therapy demonstrate how certain techniques and focuses produce cognitive and behavioural changes in the one receiving them.

BEHAVIOR THERAPY

The emphasis placed on learning by behavior therapy has allowed it to address behavioral issues via psychotherapy. One element of behavior therapy is the use of classical conditioning, which was a phenomenon discovered in an accidental manner by Ivan Pavlov. While investigating the digestive processes of dogs he had noticed changes in their reactions to food. Early on in the investigation, the dogs would begin salivating once their meal was in front of them. Eventually, the dogs began to salivate prior to the arrival of the food. Pavlolv believed they began associating certain cues that were associated with the presentation of food, such as the arrival of a food cart. Pavlov investigated this by ringing a bell prior to giving the dogs their food. While there was no response to the sounds of the bell at first, the dogs would begin to salivate at the sound of the bell only, in the absence of any food (Rehman et al., 2021)! The process of classical conditioning involves the association of some stimulus with a normally unrelated response. In the case of Pavlov's dogs, the ringing of the bells becomes associated with the presentation of food and thus the physiological responses associated with digestion.

Conditioning relates physiological responses throughout the body to the cognitive processes of the brain. Outside of therapy, conditioning has many implications. For instance, it is involved in the training of animals, and plays an important role in the tolerance of drugs (Siegel, 2001). Classical conditioning in humans is very interesting to investigate, and many peculiar phenomena have been observed as a result. For instance, when a drug with a certain taste is administered repeatedly, sometimes the taste itself is able to elicit an immune response (Rehman et al., 2021). In the 20th Century, classical conditioning was employed to address bedwetting in children. Psychologists Orval Hobart Mowrer and Molly Mowrer noticed that many of the children for whom they were serving as house parents had the issue of bedwetting. To treat this, they developed a system where the dampening of the bed associated with bedwetting would trigger a bell, causing the child to wake up.

The goal was to associate the sensation of a full bladder with that of waking up, so that the children would be able to wake up and urinate in the bathroom when needed. The first child they tested it on would not initially wake up upon the triggering of the alarm, but the urination would stop. Eventually, the bell began waking him up and he was soon doing so prior to the bell even ringing. This success was expanded to twenty nine other children at the center where the two psychologists were serving (Doroshow, 2010).

Investigation into conditioning revealed that responses generated from association are not exclusively physiological, but also behavioral (Eelen et al., 2018). This suggests that certain behaviours may be acquired as results of past events (Raypole & Legg, 2019). An example of a method of behaviour therapy that uses classical conditioning is systematic desensitization. This is a process that is often used to overcome some sort of irrational fear, a phobia. It aims to weaken the association between the stimulus and the response. For example, for someone with a fear of dogs, it would involve disrupting the association of being around dogs with the sensation of fear. By addressing the association, it aims to stop the stimulus, i.e. the dogs, from eliciting the formerly associated response, i.e. fear (Picou et al., 2020). The process involves cycling between gradual exposure to the stimulus, and relaxation techniques. They are coupled in a way such that as soon as the patient begins feeling some sort of discomfort, the stimulus is taken away while the patient can relax to resume the desensitization process (Wynn & Ursano, 2017). The goal is to repeat this procedure until the patient does not feel fear from the greatest stimulus. Another example of how classical conditioning is used in behaviour therapy is in aversion therapy. This involves the association of some sort of negative behaviour or bad habit with an unpleasant stimulus. Ideally, this intends to add some sort of negative feeling to the unwanted behavior so that the tendency to do it is reduced (Raypole & Legg, 2019). Classical conditioning and other forms of conditioning are staples of behavior therapy, and elucidate cognitive and physiological causes of behavior. In the examples discussed, the process of association has been addressed in order to induce changes in behavior.

Cognitive therapy

Cognitive therapy operates on the basis that emotions, thoughts, and behaviors are tightly linked. It follows that by addressing negative thoughts, maladaptive behaviors are also addressed. This form of therapy greatly emphasizes control and understanding of one's thoughts. It highlights overcoming harmful ways of thinking, such as "all-or-nothing" thinking. It also serves to change how the patient fundamentally looks at things, and to convince them of how certain outlooks are wrong. For instance, in cognitive therapy for obsessive-compulsive disorder, after being informed about the different types of faulty appraisals, behavioral experiments are conducted in which their beliefs can be tested (Kaczkurkin & Foa, 2015).

One of the most well known forms of psychotherapy is cognitive-behavioural therapy (CBT). CBT is used in the treatment of many psychological disorders. It involves addressing the cognitive origins for certain conditions. For instance, in the treatment of antisocial personality disorder, CBT addresses harmful core beliefs, along with an understanding of social norms and negative behaviours (Sargın et al., 2017). A study that evaluated the effectiveness of a treatment containing CBT for borderline personality disorder found that patients reported decreased levels of anxiety and fewer dysfunctional cognitions (Matusiewicz et al., 2010). Ideally, one undergoing cognitive therapy sees amelioration of maladaptive patterns and thoughts through recognizing and understanding negative cognitions and behaviors. They are also often taught techniques to help identify and overcome such cognitions outside of therapeutic sessions (Kaczkurkin & Foa, 2015). The effectiveness of CBT is discussed in greater detail in further chapters.

Humanistic and Holistic Therapy

What is stopping me from achieving my full potential and making the right decisions for myself? This is the topic of discussion for a patient receiving humanistic therapy. The emphasis placed on the symptoms and disorders by other methods of psychotherapy is supplanted by focus placed on the individual in humanistic therapy. However, humanistic therapy also utilizes elements from other types of therapy and thus can be considered a form of holistic therapy (Center for Substance Abuse Treatment, 1999a). The details concerning both these methods of therapy are discussed in greater detail in the previous chapter. As previously mentioned, the goals pursued by humanistic therapy involve changing the way the patient sees themself. The aim is to produce a difference in one's outlook such that they can make and realize goals in order to live a better life. Changing the way one sees themselves and the world in which they live inevitably leads to changes in the way that they act. Individuals who underwent humanistic therapy have shown great, persistent changes (Elliott, 2002).

Ultimately, the uniqueness of each individual suggests that a single method of treatment will not be suitable for everyone. Even when similar symptoms are observed, there may be the need to utilize different elements of psychotherapy, and approach the situation holistically. Just as treatment strategies often use medication alongside forms of therapy to treat conditions, the integration of different elements can lead to outcomes on one's holistic health. After undergoing holistic psychotherapy, one can see improvement ranging from the realm of physical and emotional pain to that of familial relations (Shafran et al., 2017)!

Conclusion

It is clear that a scientific understanding of the brain and its processes have enabled psychotherapy to develop into what it is today. From one approach, we have been able to develop treatments for conditions using our knowledge of the relationship between our physiology and our thoughts. For instance, the use of classical conditioning that was incorporated into various forms of behavior therapy. Conversely, what may seem like conversation about problems and friendly guidance to an onlooker, can have deep implications on the neural networks that make up who we are, allowing one to better their life. There are many different facets of psychotherapy, each with their own aims for how we address the issues and can augment the way we live our lives. Psychotherapy can be used to change the way one thinks and how one acts. It can be used to overcome our greatest fears and change the way we look at and act in our closest relationships.

6 The impact of Psychotherapy
Poojitha Pai

ABSTRACT

Psychotherapy has been used to aid in the management of a plethora of diseases over the years. One increasingly important way to measure its impact on an individual's life is the Quality of Life (QoL). QoL looks at the impact of an intervention within an individual's biological, psychological, and social context. On examination of the association between psychotherapy and QoL, it is evident that it helps improve QoL for individuals suffering from mental health disorders such as major depressive disorder (MDD), eating disorders, and borderline personality disorders, to name a few. There has also been evidence to show that different psychotherapeutic interventions have helped increase QoL in individuals suffering from predominantly physical chronic illnesses such as multiple sclerosis, Painful Diabetic Neuropathy (PDN), and Inflammatory Bowel Disease (IBD). It was also observed that various factors such as age of the participant, duration of psychotherapy, variability in therapist techniques, and severity and type of disease have some influence on the impact of psychotherapy. However, more research must be done in the field to ascertain the extent of their influence.

INTRODUCTION

Florence Nightingale was one of the first medical practitioners that pushed for measuring health outcomes to evaluate interventions. She set the precedent for the medical community to constantly appraise practices by looking for specific measures. Psychotherapy is not exempt from this treatment. In the previous chapters, it was found that psychotherapy has positive effects on individuals' behaviour and cognition in the course of management of various ailments. As important as these measures are in determining the effectiveness of the intervention, a holistic assessment of the impact of psychotherapy on an individual's life is also required. Illnesses rarely exist in a vacuum. This is especially relevant for mental health illnesses. By not including the patient's own perception as well as their social context, major problems can go unnoticed. Thus, holistic assessment necessitates a biopsychosocial approach i.e. one that includes an individual's cognition and behaviour and how those can then influence other aspects of their life. These other aspects can include, but are not limited to, their perception of themselves, their ability to day-to-day tasks, and their interaction with other individuals and the society. In this chapter, I aim to assess the influence of psychotherapy on the lives of individuals with mental and physical illnesses as well as extraneous factors that may influence the level of impact on individuals.

QUALITY OF LIFE (QoL)

The Quality of Life is one measure that encompasses various biopsychosocial factors that affect life satisfaction (Haraldstad et al., 2019). It is a complex concept that varies from disease to disease. For each illness or ailment, there are various questionnaires with different properties that are specific to the outcomes of the disease. For example, some properties to be assessed in Major Depressive Disorder (MDD) could be relationship status and the ability to achieve goals set for themselves (IsHak et al., 2011) while some properties for Inflammatory Bowel Disease could be perceived stress and systemic symptoms (Marinelli et al., 2019). QoL allows clinicians to see individual patients within their social context, instead of viewing them just from the disease point-of-view. Other than quantifiable measures, there are also subjective measures that can be included in the QoL. These subjective, and often qualitative measures, allow the patients to talk about possible issues they may be facing without being confined to a number scale. This allows the patient to also be included in the process of treatment evaluation and allows practitioners to also grasp otherwise unknown information. Some common tools that are used to the assess QoL in individuals include Short-Form Health Survey (SF-36), World Health Organization Quality of Life Assessment Instrument (WHOQOL-100) and Quality of Well-Being (QWB) scale – all of which encompass domains of life such as daily functioning, social relationships, and emotional regulation.

There have been numerous studies and reviews conducted on evaluating the effect of psychotherapy on the QoL of individuals suffering from a variety of illnesses. The impact on QoL for few well researched illnesses (both mental and physical) will be highlighted below.

Predominantly Mental Illnesses
Major Depressive Disorder (MDD)
It is well established that QoL is affected negatively during the course of MDD. This low QoL can be attributed to various depressive symptoms such as loss of pleasure, changes in sleep and appetite, fatigue, long-lasting periods of low moods, and even suicidal ideation (IsHak et al., 2011). All of these not only affect the individual, but also their social surroundings i.e. family and friends. In turn, the environment and social relationships can also have major effects on a person's mental state and can be important indicators of problems. The commonly used psychological interventions include Cognitive Behavioral Therapy (CBT), interpersonal therapy, and supportive therapy. CBT is a structured therapy modality that predominantly helps individuals become aware of how negative automatic thoughts, attitudes and beliefs cause feelings of sadness. Similarly, interpersonal therapy is also structured but with a higher focus on establishing and maintaining relationships. Finally, supportive therapy is a non-structured modality that relies more heavily on the therapists' interpersonal skills to encourage the individual by providing support, reflection and empathetic listening without focusing on classical therapeutic strategies (Health Quality Ontario, 2017). In

addition to psychotherapy, pharmacotherapy, i.e. medication, is also often used in isolation or combination to alleviate more biological pathologies related to MDD.

One review of 36 studies on the impact of psychotherapy, pharmacotherapy, and their combination on QoL in depression found that psychotherapy was significantly helpful in improving QoL. Some included studies, however, showed that a combination of pharmacotherapy and psychotherapy showed the highest increase in QoL (IsHak et al., 2011).

In another study conducted to look at the differences in effectiveness of the type of therapy, CBT, interpersonal therapy, and supportive psychotherapy were all able to significantly decrease the symptoms of MDD compared to the controls, with structured therapies being more effective post-treatment and at follow-up. Similar to previous literature, the highest short- and long-term improvements were seen in groups that had a combination of some form of psychotherapy and pharmacotherapy (Health Quality Ontario, 2017).

Another group conducted a meta-analysis of 44 randomized control trials comparing QoL of adults with depression with and without psychotherapy. They showed that there was a moderate but significant connection between those who received psychotherapy and a higher QoL. This evidence shows that psychotherapy is not only useful for reducing depressive symptoms but also in increasing QoL which they measured by looking at domains such as mental functioning, social and work-related relationships, and engagement in everyday life (Kolovos et al., 2016).

Eating Disorders

Eating disorders are chronic and often disabling conditions that can severely affect one's QoL. In fact, studies have shown greater QoL impairments in individuals with eating disorders compared to other mental health conditions (de la Rie et al., 2005). Many domains of an individuals' lives are affected including physical, psychological, emotional, and social spheres. Symptoms vary between each disorder but many often include a distorted perception of themselves, binge-purge behaviours that can affect normal bodily processes severely, anxiety, and their ability to maintain relationships (DeJong et al., 2013). While reduction of these symptoms is crucial, many individuals seek out help due to the impairment in QoL. For this reason, both subjective and objective measures of QoL that include their own ideation as well as objective reduction in the symptoms must be obtained in order to determine the effect of psychotherapy.

Most commonly, Cognitive-Behavioral Therapy (CBT) is used to help manage the symptoms of eating disorders. CBT has been further adapted to specifically target eating disorders, leading to the conception of variations such as CBT for bulimia nervosa (CBT-BN), CBT for binge eating disorder (CBT-BED), and Enhanced CBT (CBT-E) for eating disorders in general. These psychotherapy types aim to address the underlying psychological and behavioural mechanisms of eating behaviours. The focus here is more behavioral and helps individuals change behaviors now and in the future, aiming to have a long-lasting impact on their lives. Group-based CBT is also often used where, in addition to the therapist facilitating conversation, there are other factors involved such as interpersonal feedback, social learning, emotional expression, and group cohesion (Grenon et al., 2017).

Over the decades, these techniques have shown very promising results in reducing symptoms and behaviors related to eating disorders. It has also helped increase the QoL of individuals suffering from eating disorders. One meta-analysis of 34 articles that assessed the impact of various types of CBT on individuals with eating disorders showed that there were significant and modest improvements in QoL, when compared to other types of therapy, psychoeducation, and controls. The subjective QoL as well as the health related QoL were significantly higher in those that received CBT. Post-treatment, therapist-led CBT showed highest improvement in QoL and general recovery compared group-based CBT as well as the control. At 12 month follow-up, group based CBT showed vast improvements suggesting that group-based modalities may require more time than therapist-led ones to achieve their goals (Linardon & Brennan, 2017). Regardless of the type of CBT or psychotherapy used, there is good evidence to show that psychotherapy is effective at increasing QoL for patients with various eating disorders and that this effect still exists even after the treatment sessions are over.

Borderline Personality Disorder (BPD)

BPD is a chronic psychiatric disorder that is characterized by instability in relation to self-image and interpersonal relationships with heightened suicidal ideation and impulsivity. Individuals with BPD are associated with a marked decrease in QoL due to their symptoms and suffer high rates of mortality compared to other populations (Kulacaoglu & Kose, 2018). Psychotherapy for BPD explicitly addresses improving QoL by decreasing symptoms associated with the disease. Common forms of psychotherapy that are used to aid individuals with BPS include dialectical behavior therapy (DBT), Mentalization-based therapy (MBT), and Systems Training for Emotional Predictability and Problem Solving (STEPPS) (Chakhssi et al., 2021) among others. All these therapy modalities provide focused and active interventions that emphasize current relationships and daily functioning. These modalities include a therapist-written manual for common clinical problems that individuals may face, a generally structured format so that the individuals may develop proactivity and self-agency, a focus on emotional processing that clearly connect 'acts' with 'feelings', and a robust understanding of the treatment plan so that individuals are aware and active in their management of the disease (Kulacaoglu & Kose, 2018).

One meta-analysis analysed 14 randomized control trials that compared psychotherapy of some form for individuals with BPD with controls. These studies all showed significant increases in the QoL of individuals receiving psychotherapy. The therapy modalities not only reduced BPD symptoms, but also were associated with increases in physical, mental, social, and emotional functioning for individuals with BPD. However, in many cases, it was found that the improvements in BPD pathology severity through psychotherapy was not always correlated with improvements in QoL. This suggests that impairments in QoL may still exist even if BPD pathology severity is decreased through psychotherapy. The takeaway from this study was that BPD pathology severity and QoL increase are both inherently different constructs. Therefore, to maximize the impact of psychotherapy in individuals with BPD, both the BPD pathology as well as a QoL increase must be focused on in a multi-pronged approach (Chakhssi et al., 2021).

Chronic Diseases

In addition to aiding those with mental illnesses, psychotherapy can also have an impact on the QoL of those suffering from chronic physical illnesses. Psychotherapy can also be very important for those suffering from terminal illnesses and during bereavement and grief, which will be highlighted in later chapters. Physical illnesses do not occur in a vacuum either. Chronic diseases and impending mortality can have very severe effects on an individual's mental health, which can profoundly affect patients and those around them. In this section, the impact of psychotherapy on chronic illnesses will be evaluated in terms of QoL.

Simply having a recurring chronic illness can be a major obstacle that can significantly impair QoL in terms of mental health, self-image, functionality, and even financially. Many chronic diseases are also associated with stress and uncertainty as patients have to constantly monitor their disease progression, adjust to new lifestyles, and endure debilitating and demanding treatments (White, 2001). Some common chronic conditions include diabetes, arthritis, kidney diseases, multiple sclerosis among many others.

Different psychotherapeutic and psychosomatic modalities are used to combat chronic diseases. The main targets of these interventions are psychological symptoms, personality traits, attitudes towards disease and life, and social isolation (Deter, 2012).

For example, multiple sclerosis (MS) that is characterized by both somatic symptoms (e.g. pain and fatigue) as well emotional and psychological symptoms (e.g. problems with thinking, learning and planning, depression, anxiety) requires both biological and psychological interventions to increase QoL (Gil-González et al., 2020). One study showed that 5 weeks of group positive psychology (which included positive psychology exercises such as gratitude for positive events, enjoyable and meaningful activities, remembering past successes etc.) for individuals with MS showed improvement in QoL. However, more randomized control trials are required in this field.

Another chronic disease, Inflammatory Bowel Disease (IBD) causes many emotional stresses in addition to their physical impairments. These emotional stresses include a loss of bowel control, feeling unclean, poor body image, and social isolation. One systematic review reported that psychotherapeutic interventions such as CBT, acceptance commitment therapy, stress management programs, and mindfulness had a significant positive impact on their QoL.

One subset of these diseases includes those that cause chronic pain in individuals. Chronic pain is defined as a persistent or recurring pain that lasts longer than 3 months (Lim et al., 2018). It is a complex stressor that presents many challenges to different faculties of an individual – physical, psychological, and even financial. As a result, it necessitates a multi-pronged approach that is able to address many of these challenges. Chronic pain has been shown to negatively impact quality of life due to lower levels of functionality, higher rates of fatigue, and higher susceptibility to problematic pain medication usage. They are also more likely to suffer from other psychiatric conditions such as MDD, anxiety disorders, and post-traumatic stress disorders (Sturgeon, 2014).

Common psychological interventions for individuals with chronic pain include CBT, operant-behavior therapy (OBT), mindfulness-based therapy (MBT), and acceptance and commitment therapy (ACT). Psychotherapy for chronic pain primarily targets improvements in physical, emotional, and social functioning rather than focusing on the resolution of the pain itself.

Painful Diabetic Neuropathy (PDN) is a condition that is developed by approximately 25% of individuals suffering from type-2 diabetes mellitus (T2DM). PDN has been shown to affect mental and physical well-being, which include pain, anxiety, and depression. While pharmacotherapy aims to relieve pain, rehabilitation often includes physical therapy (i.e. exercise) and psychotherapy, often in the form of CBT. Few studies have shown that CBT had a moderate effect on pain severity, similar to other modalities such as mindfulness meditation and mindfulness-based stress reduction – both of which also increase QoL moderately in patients with PDN (Laake-Geelen et al., 2019).

From the various diseases highlighted, it is evident that psychotherapy has a positive impact on the QoL of individuals living with chronic illnesses. However, the level of impact and type of modality used differs between diseases. While more randomized control trials are required, one important takeaway is that psychotherapeutic interventions must be included in the management of chronic illnesses.

FACTORS THAT INFLUENCE THE IMPACT OF PSYCHOTHERAPY ON QOL
On close examination of the papers looking at the association between psychotherapy and QoL in various contexts, there were some factors that influence the impact that psychotherapy can have on an individual.

The first factor to discuss is the duration of psychotherapy received by the individual. One study examining the impact on QoL of individuals receiving psychotherapy for various illnesses showed that the more treatment they received, the better their QoL improved. The duration of their treatment as measured by the number of psychotherapy sessions that they attended was significantly correlated with their post-treatment QoL scores. However, no data was provided on the follow-up (Crits-Christoph et al., 2008). Another study that examined the association between psychotherapy duration and QoL found that there was no significant difference in QoL scores post-treatment in individuals who received short- and long-term psychotherapy for alcohol abuse. However, in the two years of follow up, it was shown that those who received short-term therapy were more likely to engage in higher alcohol consumption suggesting that long-term psychotherapy is more beneficial in the long run (Jyrä et al., 2017).

Another factor that influenced the impact of the therapy was the age of the participant. Younger participants were found to often not respond well to psychotherapeutic interventions. In one study examining the impact of psychotherapy and its side effect in young adults, the researchers found that the perceived lack of control around when and how to engage in therapy contributed to participants experiencing side effects of psychotherapy (Lorenz, 2021). However, the impact of psychotherapy does not increase linearly with age. One previously mentioned study that looked at the connection between psychotherapy and QoL in patients with MDD showed that elderly patients showed no significant improvements in QoL even after one year of interpersonal theory (IsHak et al., 2011). However, the reasons for this are unclear and more research must be done to expand on the associations between age and impact of psychotherapy.

Some other factors that were observed were the type of disease, gender (Crits-Christoph et al., 2008) and variability in therapists. However, these have not been fully explored and high quality studies must be conducted to understand their influence on the impact of psychotherapy.

Conclusion

The QoL has become increasingly important in measuring health outcomes regarding psychotherapy. We have been able to show that many forms of psychotherapy are helpful in improving QoL for mental health disorders such as depression, eating disorders, and personality disorders. Psychotherapy has also been very helpful in aiding those with chronic illnesses such as multiple sclerosis, IBD, and PDN, which has allowed them to attain higher QoL. Even though psychotherapy in some form has existed for over a century, there is still much to still be learned and discovered.

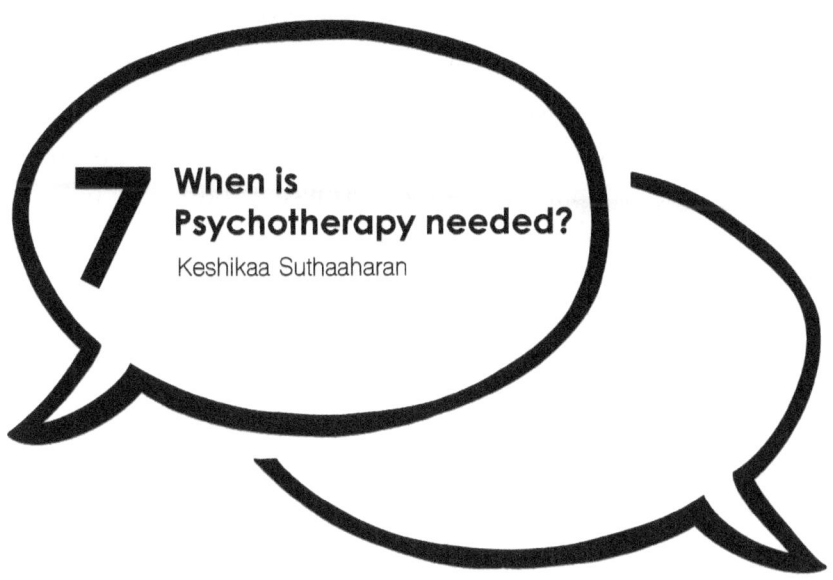

7 When is Psychotherapy needed?

Keshikaa Suthaaharan

Abstract

This chapter explores the contexts in which psychotherapy is needed. It is evident that psychotherapy can be used in a wide variety of clinical contexts and has numerous benefits to patients. In particular, psychotherapy can be used to treat specific mental health disorders, including obsessive-compulsive disorder, post-traumatic stress disorder, major depressive disorder, and generalized anxiety disorder. In addition to mental health disorders, psychotherapy can be used to help patients and families experiencing cancer treatment and palliative care. Finally, psychotherapy can help people process grief, whether it be the normal grieving process or more nuanced conditions such as complex grief. Interestingly, psychotherapeutic techniques have been adapted or developed to each specific condition. Though this chapter does not provide an extensive list of the contexts in which psychotherapy is needed, it highlights the diverse applications of the field. In addition, it indicates the potential for psychotherapeutic treatment of a wider variety of diseases. Thus, psychotherapy is a highly promising versatile tool and has shown applications within a variety of different contexts.

Introduction

As the previous chapters have shown, psychotherapy has a rich history and has developed into a prominent field. Today, the field of psychotherapy has numerous impacts on society, as well as on individual patients. One of the main impacts of psychotherapy is the improvement of the quality of life of patients affected by both predominantly mental and predominantly physical diseases, as described in Chapter 6. To build upon the concepts discussed in the previous chapter, I will discuss a few specific contexts in which psychotherapy is needed, and the development or modifications of psychotherapeutic interventions for these contexts. First, I will discuss the use of psychotherapy to treat four different mental health disorders: obsessive compulsive disorder, post-traumatic stress disorder, major depressive disorder, and generalized anxiety disorder. Following this, I will discuss other applications of psychotherapy, specifically in supporting cancer care, palliative care, and people experiencing grief.

Obsessive Compulsive Disorder

One of the key areas where psychotherapy is needed is in treating obsessive compulsive disorder (OCD). Obsessive compulsive disorder is characterized by obsessions and compulsions (International OCD Foundation, n.d.). Obsessions are intrusive thoughts, images or impulses that usually cause extreme distress. These become more and more frequent and impact the person's ability to participate in daily activities. In order to reduce the anxiety caused by obsessions, compulsions, repetitive and ritualistic behaviours or thoughts, are performed (International OCD Foundation, n.d.). Compulsions are also often performed to prevent something bad from happening (Anxiety Canada, n.d.). These are often time-consuming and can greatly impact a person's daily life (International OCD Foundation, n.d.)

In treating OCD, psychotherapy is aimed at reducing the severity and frequency of the obsessions and compulsions common to the disorder. Cognitive behavioural therapy (CBT) is commonly used to treat OCD. CBT, in general terms, relies on modifying behaviours and cognitions, and providing people with new positive experiences to override previous negative experiences (Hazlett-Stevens & Craske, 2002). A common CBT method utilized in treating OCD is exposure and ritual prevention (E/RP), which focuses on helping patients relieve symptoms without performing compulsions (Foa et al., 2012). This method consists of four main components, any combination of which will be used within a patient's particular treatment program: in vivo exposure, imaginal exposure, ritual prevention, and processing (Foa, 2010). Firstly, in vivo exposure is based on directly exposing patients to an object or situation that triggers obsessional distress and compulsions (Foa et al., 2012). This is used to help a patient confront the feared stimuli (Foa, 2010). Imaginal exposure, unlike in vivo exposure, relies on patients imagining feared stimuli in order to help patients work through the consequences they believe will occur if compulsions are not performed (Foa, 2010). Ritual prevention, as the name suggests, encourages patients to not perform the compulsions associated with addressing their obsessions (Foa, 2010). Processing is the reflection component of E/RP, allowing the patient to think about whether not performing the compulsions led to the outcomes they expected (Foa, 2010).

Through utilizing some or all of these components, E/RP is needed to treat the obsessions and compulsions of OCD, and to help patients develop behaviours that can reduce fear and anxiety. In this way, psychotherapy is a highly beneficial tool for supporting patients with OCD.

POST-TRAUMATIC STRESS DISORDER

Multiple psychotherapy techniques are needed to treat post-traumatic stress disorder (PTSD). PTSD is a mental health disorder caused by being exposed to or witnessing a traumatic event such as a natural disaster, a serious accident, or combat (American Psychiatric Association, n.d.-b). Patients with PTSD experience flashbacks to the traumatic event, feel intense emotions related to the event, avoid situations that remind them of the event, and react strongly to innocuous stimuli. PTSD can cause changes in arousal and reactivity, including emotional outbursts, and can co-occur with disorders such as depression (American Psychiatric Association, n.d.-b).

In PTSD, maladaptive psychobiological stress responses lead to the numerous symptoms present (Bonne et al., 2004). Thus, PTSD treatments primarily focus on two main outcomes: 1) reducing the psychobiological stress responses that leads to distress, and 2) reducing maladaptive psychobiological processes such as sleep disturbances, depression and anxiety (Wilson et al., 2012). Key mechanisms of recovering from a trauma, as reviewed in Wilson et al. (2012), include: emotional engagement with traumatic memories, organization of traumatic memories, and changing core beliefs about the world and self. Psychotherapy can utilize these key mechanisms to promote recovery from PTSD (Wilson et al., 2012).

Due to the large role of trauma in PTSD, psychotherapy treatments are often trauma-focused, in other words, directly focusing on the experience and thoughts of the traumatic event (Watkins et al., 2018). A couple of cognitive behavioural therapy (CBT) interventions are trauma-focused. CBT, in the context of PTSD, CBT focuses on helping patients identify and change irrational thoughts about the traumatic event, and modify harmful behaviours as well (Watkins et al., 2018). Three common CBT treatments used to treat PTSD are prolonged exposure (PE), cognitive processing therapy (CPT), and cognitive therapy. Each of these can address different aspects of the traumatic event and its effects. PE, which utilizes a combination of psychoeducation, in vivo exposure, imaginal exposure and breathing retraining, can help a person emotionally process the traumatic event and the fears surrounding it (Watkins et al., 2018). Cognitive processing therapy can help patients process how the traumatic event has affected their thoughts about themselves, others and the world, and further identify and challenge maladaptive cognitions (ex. "It was my fault that it happened"). Cognitive therapy is focused on the identification of memory attributes and triggers, as well as cognitive and behavioural practices, that contribute to PTSD symptoms (Ehlers et al., 2005). Cognitive therapy can also help patients change their negative evaluation of the traumatic event and the events occurring directly afterwards (Ehlers et al., 2005). In addition to CBT, eye movement desensitization and reprocessing (EMDR), a psychotherapy technique that utilizes rapid eye movements, is used to process various aspects of the traumatic memory (Schnurr, 2017).

Alternatively to trauma-focused treatments, non-trauma-focused psychotherapies can be used as well. For example, Present-Centered Therapy (PCT) is centered on changing present behaviours, developing a patient's understanding of how trauma can impact their life, and educating about problem-solving strategies (Frost et al., 2014). Interpersonal psychotherapy (IP), which focuses on interpersonal experiences, is also highly useful because PTSD individuals often develop a mistrust of others or alienate themselves due to their traumatic experiences (Markowitz et al., 2015).

PTSD is a disorder that is caused by a traumatic event. Psychotherapeutic treatments ultimately focus on reducing the effect of a traumatic event on the affected person's life. Trauma-focused treatments such as specific CBT interventions and EMDR can help PTSD patients both come to terms with the traumatic event and challenge cognitions that cause distress. As alternatives to trauma-focused treatments, non-trauma focused treatments such as PCT and IP also have notable benefits for PTSD patients. A combination of various psychotherapeutic techniques can help patients effectively work through their PTSD.

Major Depressive Disorder

Major depressive disorder, hereafter referred to as depression, is a serious mental health disorder that impacts a person's thoughts, feelings and behaviours (American Psychiatric Association, n.d.-a). Symptoms of depression vary, but common symptoms include feelings of sadness, changes to sleep patterns, changes in appetite and thoughts of suicide (American Psychiatric Association, n.d.-a) Depression can also affect motivation, cognition, and induce multiple physiological changes (Gilbert, 2007).

For patients with depression, a number of psychotherapy techniques can be utilized to target different symptoms of depression. CBT, described earlier in this chapter, can be used to address the underlying issues of depression, discuss the impact of situational stressors that a patient may be experiencing, and much more (Gautam et al., 2020). CBT for depression is focused on identifying major problems, setting goals, modifying thoughts and behaviours, completing homework related to developing new skills and strategies, and preventing relapse (Gautam et al., 2020). In addition to CBT, behavioural activation therapy (BAT) is also used for treating depression, and this intervention focuses on helping patients identify pleasurable activities and incorporate them more into their lives (Cuijpers et al., 2019). Interpersonal psychotherapy, which focuses on understanding the interaction between interpersonal relationships and depression, has also shown effectiveness (Cuijpers et al., 2019; Klerman & Weissman, 1994). Finally, problem-solving psychotherapy, focused on developing skills in problem-solving, is also useful in treating depressive symptoms (Nezu, 1986).

Depression is a complex mental health disorder that causes numerous physiological and cognitive symptoms. Psychotherapy can be used effectively to address many of these symptoms, with CBT, BAT, interpersonal therapy and problem-solving therapy targeting different aspects. Ultimately, psychotherapy can greatly improve the lives of those living with depression.

GENERALIZED ANXIETY DISORDER
Generalized anxiety disorder (GAD) is a mental health disorder characterized by excessive and persistent worrying (Anxiety and Depression Association of America, n.d.). People with GAD will worry about any number of things, including school, work, and health; this worry may be about real-life situations or about potential future catastrophes (Anxiety and Depression Association of America, n.d.). Worry in GAD patients can be thought of being caused by a general state of uncertainty and the inability to tolerate it (Simos & Hofmann, 2013).

Psychotherapy is needed to reduce the worrying associated with this disorder. CBT is one of the most common psychotherapy interventions used to treat GAD. This technique typically begins with psychoeducation, in which patients are educated about the disorder and what it means for their lives (Simos & Hofmann, 2013). Self-monitoring is an important part of CBT for GAD, and involves recording the intensity, duration and frequency of their worries (Simon et al., 2020). In addition, since patients with GAD often underestimate their coping abilities and inflate the probability of negative events, CBT is used to restructure and challenge these thoughts (Simon et al., 2020). Relaxation exercises, worry exposure (getting patients to purposefully think about their worries) and worry behaviour control (encouraging patients to stop using behaviours to avoid their worry) are also commonly used in CBT (Simon et al., 2020). Problem-solving strategies are taught to patients to help them resolve GAD-associated problems, particularly the intolerance of uncertainty (Simon et al., 2020). Imaginal exposure may also be used to process any core fears held by the patient (Simos & Hofmann, 2013).

Psychotherapy as a treatment for GAD plays an important role in helping patients process their worries. In addition, it provides them with the ability to find solutions to worrying situations when they do arise. Ultimately, psychotherapy provides GAD patients with ways to change cognitions about themselves and external situations.

CANCER CARE

Along with treating mental health disorders, psychotherapy is also needed within cancer care. Cancer is a prevalent disease within Canada, with an estimated 1 in 2 people being diagnosed with cancer, and 1 in 4 people dying from the disease in their lifetime (Government of Canada, 2016). During treatment, cancer patients often face a variety of physical symptoms including fatigue, cognitive impairment and pain (Adler et al., 2008a). Importantly, however, patients also face psychological issues, such as experiencing anxiety and depression due to stressors caused by the illness (Adler et al., 2008b). This unique combination of physical and mental needs requires the need for an integrated health care approach. Despite this, while there have been incredible advances from a medical perspective, the psychological needs of patients are often not met in the clinical setting (Adler et al., 2008b).

Recent years have shown the emergence of psychological therapies that can improve quality of life, despite the stress caused by the illness and by cancer treatments. Supportive psychotherapy is a therapeutic intervention that provides support as needed during the patient's illness (Watson & Kissane, 2011). This intervention is aimed at facilitating the processing of strong emotions and promoting effective coping with their illness (Watson & Kissane, 2011). Supportive psychotherapy is ultimately a flexible intervention that utilizes elements of cognitive behavioural therapy and psychodynamic therapy (Adler et al., 2008a). In addition to supportive psychotherapy, cognitive behavioural therapy is also utilized in cancer care to improve psychological symptoms as well as the physiological symptoms of treatments (Watson & Kissane, 2011).

Apart from the cancer patients themselves, psychotherapy is also often used to support the families and caregivers of patients. A loved one's diagnosis can lead to feelings of distress, coupled with navigating the new responsibility of caring for the loved one (Sherman & Simonton, 1999). Psychotherapeutic interventions such as cognitive behavioural therapy can decrease stress, anxiety and depression faced by family caregivers (Borji et al., 2017).

Within cancer care, psychotherapy is essential in supporting the emotional and mental wellbeing of patients and families. Furthermore, psychotherapy can help to reduce the discomfort and physical symptoms caused by cancer treatment. This highlights that psychotherapy is needed within an integrated health care approach, one that meets the complex needs of cancer patients.

PALLIATIVE CARE

Palliative care refers to the holistic model of healthcare that treats those with life-threatening health-related conditions, particularly those at the end-of-life (Radbruch et al., 2020). Palliative care focuses on enhancing the quality of life of patients with life-threatening conditions and their families (Radbruch et al., 2020). Palliative care patients often face challenges such as coping with their condition, and making preparations for their death (Gramm et al., 2020). In addition, palliative care patients often face psychological issues such as despair, anxiety, hopelessness, loss of meaning in life, and loneliness (Gramm et al., 2020).

Psychotherapy can support patients through many of the aforementioned challenges. For example, psychotherapy can be used to confront emotions surrounding death (Vachon, 1988). Psychotherapy has also been shown to reduce anxiety and depression symptoms, and increase patients' quality of life (Fulton et al., 2018). The need for meaning and spirituality within cancer patients can also be addressed through psychotherapy. In particular, meaning-centered psychotherapy (MCP) was developed to help terminal cancer patients maintain a sense of purpose and meaning in their lives, and simultaneously diminish despair about dying (Breitbart, 2016; Breitbart & Duva, 2016). MCP has been shown to be effective in decreasing hopelessness, depression, anxiety, and other issues faced by palliative care patients (Breitbart, 2016). Another example of psychotherapy applied to the palliative care context is cognitive-existential group psychotherapy (CEGP). CEGP focuses on six goals centered on aspects such as grief, problem-solving, coping, modifying negative thoughts, promoting hope, and more (Gramm et al., 2020).

Similar to within cancer care, families of palliative care patients can also be supported by psychotherapy. Within palliative care, family members face major stressors namely, 1) coping with the illness of their loved one, and 2) adjusting to their new role as informal caregivers for their loved one (Kristjanson & Aoun, 2004). Existential issues, such as coping with increased awareness of mortality and with feelings of loneliness and isolation, are also key burdens faced by caregivers (Applebaum et al., 2015; Fegg et al., 2013). Existential behavioural therapy was developed in 2013 to support informal caregivers experiencing existential issues (Fegg et al., 2013). This intervention is focused on facilitating discussion surrounding death, finding meaning, self-care, bereavement, stress management, mindfulness, saying goodbye to their loved ones, and next steps (Fegg et al., 2013).

Within palliative care, psychotherapy is used in a variety of settings, from supporting patients to supporting patients' relatives. For patients, this relief can help patients achieve meaning despite being at the end-of-life. For relatives, psychotherapy can relieve the burden of being a caregiver and improve coping strategies. Ultimately, in palliative care, psychotherapy is needed as an essential part of the healthcare model to reduce distress and improve the quality of life of everyone involved.

GRIEF

Psychotherapy is often needed to help support people experiencing the loss of someone in their lives and to facilitate the healing process. Several cognitions can interrupt the grieving process and prevent healing. In particular, there can be a tendency for rumination, that is, thinking repetitively about the circumstances surrounding the death (Kosminsky, 2017). Other maladaptive cognitions are also common. Cognitive behavioural therapy can help patients challenge these cognitions, and further focus on positive memories with the loved one (Kosminsky, 2017). In the case where the deceased is someone that the patient fears or that they did not have the best relationship with, psychotherapy can still help navigate the complicated feelings surrounding grief (Neimeyer, 2012).

The death of a loved one leads to acute grief, which is characterized by thoughts of the deceased, disruptions to interpersonal relationships, and a perceived loss of meaning in life (Shear, 2012). Acute grief typically subsides into integrated grief, where grief becomes integrated into the background, and the affected person resumes normal activities within day-to-day life (Shear, 2012). However, some people will instead experience complex grief (CG), where acute grief symptoms become prolonged and impact life to a greater degree (Shear, 2012; Shear et al., 2013). When experiencing CG, people are unable to progress to integrated grief, and so the healing process is interrupted (Shear et al., 2013). Psychotherapy can assist in navigating CG; complex grief psychotherapy (CGP) was recently developed for this purpose. Many of the techniques of this intervention are those used in prolonged exposure, described in a previous section of this chapter (Wetherell, 2012). This type of psychotherapy is also goal- and relationship-focused (Wetherell, 2012). CGP is ultimately focused on removing any barriers to integrated grief that the patient faces, and directing the grieving process towards acceptance of the death of their loved one (Wetherell, 2012). In treating CG, it can also be helpful to utilize a relational group process, where people can heal by talking with others experiencing similar grief, or to use the "empty chair" technique, where people can envision saying things to the deceased (Erskine, 2014).

Psychotherapy can greatly facilitate the grieving process, and can be adapted as needed, making it a versatile tool. In particular, psychotherapy can be used to challenge the maladaptive cognitions that stem from a death of a loved one, or to address complex grief and its effects on life. While the specific approach varies depending on the person's situation, psychotherapy is ultimately needed to help patients heal from and work through grief.

Conclusion

As explained in this chapter, psychotherapy has multiple applications in clinical settings. In particular, psychotherapy can be used to treat various mental health disorders, particularly in relieving symptoms and challenging maladaptive cognitions and behaviours. Outside of directly treating diagnosable disorders, psychotherapy can support patients within palliative care, patients undergoing cancer care, and patients experiencing grief. To account for the various uses of psychotherapy and varying patient needs, numerous psychotherapy techniques are utilized for specific contexts. For example, CBT is used to treat both PTSD and major depressive disorder, but with different goals and procedures. In other contexts, new psychotherapeutic interventions have been developed, such as complex grief therapy for supporting patients experiencing complex grief. This chapter, though not an extensive review of when psychotherapy is needed, shows that the applicability of psychotherapy in the clinical setting is amazingly diverse and that psychotherapeutic interventions are highly versatile. As the field of psychotherapy continues to grow, psychotherapy may be used in additional settings as well. The next chapter will build upon the concepts discussed in this chapter to examine whether or not psychotherapy is effective as a treatment.

8 Is Psychotherapy effective? Does it have adverse effects?
Annilea Purser

Abstract
This chapter concerns itself with the theoretical query of how effective psychotherapy is as a discipline and form of treatment for mental illnesses and disorders. Rather than looking at more minor effects such as adverse reactions, this chapter takes a step back to look at the overall consensus about whether or not psychotherapy is effective according to the scholarly community who concern themselves with this issue. This analysis does not argue one side of the debate over the other but explores opinions expressed on either side of the argument.

Introduction
Over the previous seven chapters, we have explored the field of psychotherapy, the different types and practitioners of psychotherapy, the cognitive/behavioural effects of psychotherapy, and even how it is utilized and assessed for various psychological disorders or illnesses. At this point in this discussion, the big-picture question of whether or not psychotherapy is effective in treating psychological conditions and diseases demands to be posted. However, like many queries in the field, the answer to this question is not simplistic in nature but rather requires an in-depth analysis of professional opinions, studies, and arguments

that have been made since the founding of psychotherapy by Sigmund Freud in the 1890s. This chapter will respond to the question of whether or not psychotherapy is effective, not by arguing one side of the debate over the other, but by looking at both sides (either for or against it) and contextualizing these opinions in the circumstances, some of which have been aforementioned in Chapter 5 (What are the cognitive/behavioural effects of receiving Psychotherapy?). Consequently, this chapter will explore the issue of the effectiveness of psychotherapy in a broader, more holistic nature. This will be done through a comprehensive analysis of the scholarly works and criticisms put forth by researchers on both sides of the debate, many of which are concerned with meta-analytically looking at previous research that is being done in the field.

However, before delving into the debate over the effectiveness of psychotherapy, it is important to first briefly define what will be considered as a baseline for "effectiveness." As will be discussed, there is some contention surrounding methodology for measuring psychotherapy impacts. However, for this chapter, anytime that it is not otherwise specified, 'effectiveness' in relation to psychotherapy can be understood as the percentage of occurrences in which psychotherapy has some positively associated benefits. That is, when this discussion generalizes the term "effectiveness," it is not considering traits such as longevity (which will be discussed in further detail in some specific points) but rather utilizing a baseline of some positively associated benefits. These positively associated benefits are considered, in most cases, in relation to no treatment (NT). In addition to this, it is essential to note that this chapter will be focusing on effectiveness rather than efficacy. These two terms of measurement are often interchanged without notice. However, efficacy (which will not be measured) focuses on "methodological efforts to maximize the internal validity of a study" (Hunsley et al., 2013, p.5), whereas effectiveness deals with external validity (Hunsley et al., 2013, p. 5). Furthermore, when the term 'psychotherapy is utilized in general means, it can be assumed that the entirety of the practice of psychotherapy (including all sub-categories) is being referred to. If one is interested in the smaller-scale influence of psychotherapy, including potential adverse effects, Chapters 4, 5, and 6 explore these issues in great detail.

Chapter 8

METHODOLOGIES AND MEASUREMENTS

When performing any research in the field of psychology, either one or conjunction of multiple research methods are deployed to achieve a conclusion. There are five essential research methods that are typically utilized (Best Degree Programs, n.d.);

- Case studies where qualitative research methods are undertaken to obtain an observation over a specified length of time;
- Experiments where a specific procedure is utilized with the presence of a control group, control variables, and consistent terms of measurement;
- Observational studies where qualitative research methods are being used to carefully record a given situation;
- Surveys where self-reported data is analyzed, typically in a quantitative fashion, and;
- Content Analyses where previous research reports or forms of media are observed for trends or general insight

These five methods, as one can easily observe, are quite broad and vague in nature and, since the beginning of the 21st century, research pertaining to psychotherapy has typically been a mixture of these five methodologies. However, when specifically discussing the effectiveness of psychotherapy treatment in the present day, researchers tend to rely on meta-analyses (statistical analyses which draw from multiple previously-made conclusions) of smaller studies that used a variety of research methods to determine a conclusion as to whether or not psychotherapy is truly effective. An example of a meta-analytical research study could include the work of Klaus Lieb of the University of Mainz, who performed a review of 95 different (and smaller) psychotherapy studies to understand if psychotherapy really works or if previous findings were too general (Fradera, 2017). The use of meta-analytical analysis is important to this discussion as it points to the nature of the present-day debate; rather than looking at small-scale studies on psychotherapy, researchers are concerned with the percentages of reportings showcase the effectiveness of psychotherapy as these can answer the question of whether or not it is effective once and for all.

ARGUMENTS AGAINST PSYCHOTHERAPY

Criticisms towards the effectiveness of psychotherapy can be dated back to 1952 when Hans Eysenck presented research suggesting that psychotherapy is not an effective form of treatment, but rather that it is comparable to spontaneous improvement amongst patients (McNeilly & Howard, 2008, p. 74). There have been multiple psychological research reports since the age of Eysenck where researchers have looked into the effectiveness of psychotherapy and reported rejecting the notion that psychotherapy is an effective form of psychological treatment. Arguably, the most well-known of these researchers and their efforts include Cuijpers, Karyotaki, Reijnders, and Ebert's report "Was Eysenck right after all? A reassessment of the effects of psychotherapy for adult depression" (2018), and Dragioti, Karathanos, Girdle, and Evangelou's report "Does psychotherapy work? An umbrella review of meta-analyses of randomized controlled trials" (2017). Beginning with the first research paper by Cuijpers et al., the group analyzed 369' effects' or prior studies with the purpose of re-examining Eysenck's opinion that psychotherapy is ineffective (Munder et al., 2018, p.269). More specifically, the group meta-analytically examined the 369 studies by comparing an intervening variable of adults with depression to a control group of those treated with psychotherapy (Munder et al., 2018, p.269). They then further checked for four out of the six types of bias that are outlined in the Cochrane Collaboration's tool for assessing the risk of bias in randomised trials (Munder et al., 2018, p.269). Professor Cuijpers and his associates found that a significant standardised mean difference (SMD) of 0.70 existed between the treated and untreated patients, which indicates that psychotherapy is, in fact, very effective in treating depression (Munder et al., 2018, p.269). However, following their test for biases, the group concluded that the 'true' or 'proveable' SMD would most likely be between 0.2 and 0.3, indicating that there is only a small chance of psychotherapy being effective (Munder et al., 2018, p.269). The group found this conclusion on the basis that psychotherapists tend to report bias findings and that spontaneous recovery or other intervening variables were not tested (Munder et al., 2018, p.269). Therefore, although Cuijpers and his associates discovered an SMD of 0.70 initially, the lack of circumstantial evidence and potential biases resulted in them concluding that psychotherapy is more likely than not ineffective (Munder et al., 2018, p.269).

The research of Dragioti and their associates, Karathanos, Girdle, and Evangelou, echoed the research of Professor Cuijpers and his team. However, Dragioti and their team were able to expand on the analysis of Cuijpers and look beyond just depressive disorder, towards other mental illnesses and disorders. Through a series of meta-analyses of existing literature drawing on over 5,000 studies, Dragioti et al. found that various types of psychotherapy are 80% effective in treating mental illness (Dragioti et al., 2017). However, similar to Cuijpers, once they placed this data through bias checking, very few, only about 7%, were found to be 100% sound analyses (Dragioti et al., 2017). Of this 7%, the researchers discovered that Cognitive Behavioural Therapy had six conclusive and non-bias studies showing that it is effective, Mediation therapy had one, cognitive remediation had one, counseling had one, and mixed types had 7 (Dragioti et al., 2017). Thus, the research of Dragioti et al. agreed with the research of Cujipers et al. and expanded on it by including other mental illnesses and categorizing the various types of psychotherapy treatments.

The criticisms of psychotherapy studies as expressed by Cuijpers et al., and Dragioti et al., in their meta-analytical studies share many similarities, with the most prominent being the bias of researchers. As aforementioned, in both studies, the researchers discovered that when observing case studies from an overarching lens, it appears as though psychotherapy is an effective form of treatment, but when these findings are tested against bias standards, there is limited evidence supporting those claims (Jarrett, 2017). This issue of bias control is one which is extremely complex and multi-faceted. Firstly, both research groups identify that the high-level of bias is motivated by the significant heterogeneity between existing trials and their methodologies (Jarrett, 2017). That is, researchers studying the effectiveness of psychotherapy often find similar results due to similar methodologies, which has created a predominant trend for positive contributions on the topic, resulting in a tendency for researchers to only publish positive results (Jarrett, 2017). Secondly, there remains a small-study bias where researchers perform limited-capacity case studies and generalize the findings as being absolute (Jarrett, 2017). Thirdly, outsider critics like Alex Fradera express that undeclared researcher bias greatly affects the reliability of findings (Jarrett, 2017). As Klaus Lieb of the University of Mainz found by examining 95 studies on psychological treatments, scholarly research journals continue to lack taking appropriate measures to report non-financial conflict of interests (Fradera, 2017). For example, journals and authors often do not report a researcher's field of practice which is problematic as "psychotherapy researchers who realise that the effect of the therapy to which they are allied is less beneficial than another therapy cannot easily switch their research programme to another therapy (since they have often been trained in that therapy for many years)" (Fradera, 2017). In this case, this lack of conflict of interest reporting could mean that a doctor who only deals with psychotherapy publishing a positive report to support their job (and livelihood) would go unflagged (Fradera, 2017). Overall, understanding these specific biases that are pointed to in the conclusions of Cujipers and Dragioti is critical to understanding the argument that psychotherapy is not effective.

Following these findings, multiple research methodology fanatics and psychologist-researchers have suggested ways in which this issue of bias in reporting related to psychotherapy can be remedied. To counter the issue of bias, one suggestion is to look at observational findings instead of case studies as observational data is far less controlled in a hands-on manner than case studies (Jarrett, 2017). A suggestion to deal with the issue of reporting bias from authors and journals is to promote an increased amount of research findings related to negative findings or hypothesis tests to becoming published as this will diversify the field and lessen pressure for researchers to constantly identify positive conclusions (Jarrett, 2017). Further, critics state that journals should become more efficient in testing for non-financial biases and apply the same amount of rigour to assessing non-financial biases that they do for financial biases (Jarrett, 2017). Importantly, however, there is a commonly shared perspective amongst the critics (Cuijpers and Dragioti) that with the implementation of these suggestions to enhance the quality of research on the effectiveness of psychotherapy, that stronger conclusions can be drawn, even ones that are indicative of the effectiveness of it (Jarrett, 2017).

Chapter 8

ARGUMENTS FOR PSYCHOTHERAPY

On the opposite side of the spectrum of research on psychotherapy, there is a group of researchers who reject the criticisms of their counterparts and ultimately believe that existing evidence on the effectiveness of psychotherapy is conclusive in showing that it is positively effective. For the purposes of this analysis, one can turn toward the work of Abbass and his associates in the article "Is Psychotherapy Effective? A re-analysis of treatments for depression," which was released in 2018. This article is arguably one of the most effective pieces of literature in conveying the ideology of pro-psychotherapy researchers as it performs its own analysis while simultaneously addressing the concerns of opponents like Cuijpers. To begin their work, Abbass and his colleagues brought forth the work of Mary Lee Smith and Gene Glass, which found that a 0.70 SMD existed between treated and untreated patients, showing the largely positive effect of psychotherapy (Munder et al., 2018). This work was disputed by Eysenck, who brought forth the aforementioned criticisms of issues with bias and heterogeneity in the meta-analyzed cases (Munder et al., 2018).Then, Abbass discussed the work of Cuijpers, again, acknowledging his findings of lack of bias control resulting in his conclusion that psychotherapy is an ineffective form of treatment (Munder et al., 2018).

However, Abbass argues that the analyses of researchers like Eysenck and Cuijpers should not dismiss the 0.70 SMD as being illegitimate on the basis of their potential bias (Munder et al., 2018). Abbass argues that the conclusions made by opponents overly reduce the number of studies, including an overly reduced number of studies on the basis of 'possible systematic errors' (Munder et al., 2018). In this case, Abbass says that every study in psychology may deal with possible systematic errors, and dismissing them altogether is dangerous as there is a large probability that the errors are insignificant to the conclusions drawn (Munder et al., 2018). In addition, Abbass argues that the researchers utilized an overly narrow definition of risk of bias which meant that many influential, and even highly-esteemed, studies were dismissed as having 'bias' (Munder et al., 2018). For example, Abbass says that the work of Cuijpers excluded the NIMH Treatments of Depression Collaborative Research Program, even though it was considered to be the most esteemed and rigorous trial of psychotherapy to ever be conducted (Munder et al., 2018).

In addition to these criticisms that Abbass holds towards opponents of psychotherapy in relation to bias control, he also argues that the overall methodologies utilized by researchers like Cuijpers are flawed. For one, Abbass argues that, despite removing rigorous or esteemed trials on the basis of potential bias, Cuijpers and his associates included studies with significant methodological shortcomings in their meta-analysis (Munder et al., 2018). Further, Abbass argues that opponents have failed to provide significant proof of reliable methodology in their meta-analyses which reject psychotherapy as being effective, even sometimes meaning that their coding procedures or ethics processes are not recorded in any way, despite these processes being critical to ensuring the validity of a study (Munder et al., 2018).

After responding to opponents, Abbass and his associates performed their own research utilizing the same sources as Cuijpers. They removed outliers, corrected for any publication biases, restricted analysis to psychotherapy studies and utilized control groups to ultimately draw the opinion that Cuijpers should have found psychotherapy to be demonstrably effective in treating depression (Munder et al., 2018). Thus, Abbass rejects the findings of opponents on the basis of removal of too many studies and methodological errors, and instead, on the basis of his own analysis, argues that psychotherapy is an effective form of treatment (Munder et al., 2018).

Conclusion
Evidently, both sides of the spectrum of the debate of whether or not psychotherapy is an effective form of treatment for mental disorders and illnesses offer compelling arguments. Forming an opinion based on conclusions made by various researchers proves to be a difficult task, especially considering most meta-analysis conclusions discussed here are dependent on methodological preference. However, the question of "is psychotherapy effective" remains as an important question, especially with it being a popularized practice in the present day. This question

continues to be researched and debated, and in the future may lead to more conclusive and agreeable findings.

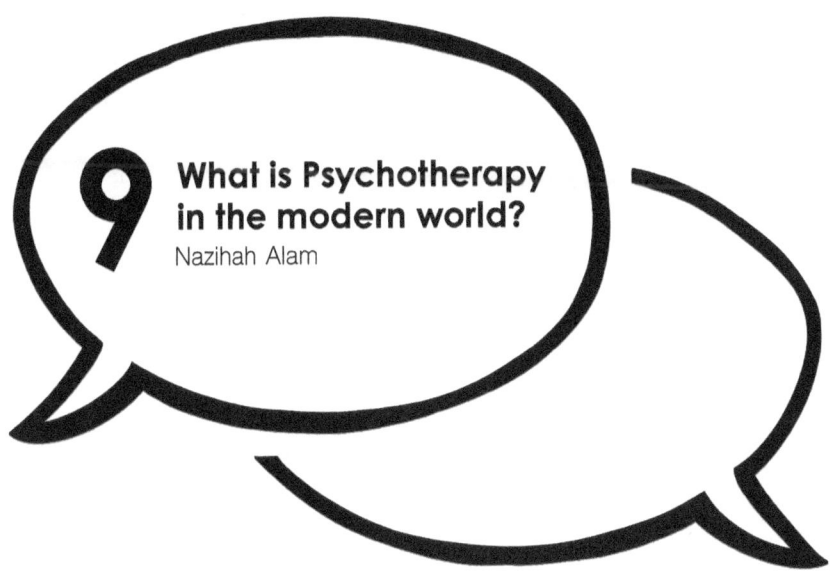

9 What is Psychotherapy in the modern world?
Nazihah Alam

Abstract
There have been many methods of treating psychological orders over the years and one of those most recent methods is psychotherapy. This chapter will be talking about how psychotherapy is in the present day. Psychotherapy in the present day is a highly collaborative nature between the individual and the therapist. It also looks briefly at the steps within psychotherapy. There are also some comparisons made between modern therapy and traditional therapy. Lastly, there is a brief look at psychotherapy used in different parts of the world.

Introduction - Brief Look at History of Treatment Used for Psychological Disorder

The treatment of psychological disorders has been advancing rapidly with the continuous research being done and the improvement of technology. As mentioned in previous chapters, there was a lot of stigma and association of madness/craziness about people with psychological disorders. One of the earliest methods to treat someone with psychological disorders was in 5000 BCE and it was called trepanation, where the patient with the illness would have a hole drilled in to figure out what is wrong (Altman et al., 2017). In the eighteenth century, there was a rise in placing people with mental illness in institutions, where the patient would be isolated and placed through different treatments (Altman et al., 2017). Moving forward there was more research development in the field of pharmacotherapy, which is the use of prescription medication as treatment. More recently, psychotherapy has begun to take over the field in modern day.

Psychotherapy in the Modern World Today

In present time, psychotherapy is most often conducted by psychologists, who are people with a Ph.D degree and are able to assess and diagnose psychological disorders but not prescribe medication (Altman et al., 2017). Psychotherapy shifts away from the standard usage of medication as a form of treatment and rather looks at a psychological treatment which is done through the method of talk therapy (Altman et al., 2017). This new approach to therapy allows an individual to explore and figure out their problematic thoughts and then find a method to overcome it and thrive (Bishop, 2020).

A very defining characteristic of psychotherapy in the modern world is that it is a very collaborative process, where both the person in therapy and the therapist form a strong bond. This can be seen as a difference from past traditional therapy, for example when looking at the period of instiualization. At that time patients were forcibly isolated and had to go through extreme treatments, such as shock therapy. There was not as much communication formulated between the individual and the person providing care. In psychotherapy, that collaborative nature is upheld by the therapist working towards a type of session during and outside of the time that is fit to help the individual (Bishop, 2020). Psychotherapy in the modern world focuses on equipping an individual with skills that can help them overcome different kinds of challenges, which can span from being more accepting of oneself or being more accepting of those people the individual may be around. These developed skills can help an individual deal with highly stressful periods of time and also help them figure out how to tolerate difficult emotions. These skills in the modern day are developed through retraining one's brain, this way certain responses that may be habitual could be worked to improve on. For example, during the covid-19 pandemic, a person may be feeling a lot of negative emotions about themselves since they are unable to go outdoors and may be constantly eating. A psychologist providing psychotherapy might help the individual perceive this situation differently. Instead of being upset about one's body, they could be encouraged to look at all the amazing things the body is doing in order to keep the person safe during the pandemic. Having this change in perspective could create all the difference, and this only one aspect of psychotherapy in the present day. As mentioned before, psychotherapy sessions tend to meet the needs of the individual and so the sessions can vary from person to person (Bishop, 2020).

Oftentimes, the time limit for each session depends and the structure of the session might not be as obvious, but the therapist in charge will try and maintain a certain direction of conversation so it does not go all over the place (Bishop, 2020). The session will usually begin with some introductions and brief talks about what is happening in between the sessions, developing a goal for the current session and figuring what topics the individual and therapist will be speaking on that day and then developing those skills to help manage the situation the individual is facing, making more goals for upcoming sessions and then concluding the session (Bishop, 2020).

As mentioned in previous chapters, the way psychotherapy is provided can differ and be provided in many forms. Some of which are group therapy where one therapist can work with a group of individuals who may be experiencing similar experiences, for example a gaming addiction (Altman et al., 2017). There is also family or couple's therapy, which looks at exploring the feelings and thoughts of a group of closely knit people in their specific relationship (Altman et al., 2017). There are also the one-to-one sessions that have been mentioned throughout this chapter, which can be effective for an individual who is working on their relationship with themselves and those around them.

PSYCHOTHERAPY IN DIFFERENT PARTS OF THE WORLD
All around the world, there have been actions to figure out ways to

break the stigma surrounding therapy. The National Health Service in the United Kingdom allows people to have free access to psychotherapy, however with that, there can be long waiting lists of people (Pajer, 2019). In Mexico, like many countries, there is still stigma on going to therapy, but the conversation has started. In bigger cities there are more affordable areas that provide psychotherapy in comparison to rural areas (Pajer, 2019). Within Canada, more people are open to the idea of reaching out to private or public therapy systems with the healthcare plans made for each province (Pajer, 2019).

Conclusion

In the present time, the use of psychotherapy to treat psychological disorders has become more prominent. As the topic of mental health is growing, more is being done to break down the stigma surrounding it. Mental wellbeing is starting to be seen as important as physical wellbeing. Throughout each stage of an individual's life, there are challenges and depending on the person, each has their own way of handling it. Although psychotherapy is a promising method, it may not be for everyone, nonetheless it is still a growing field in research today.

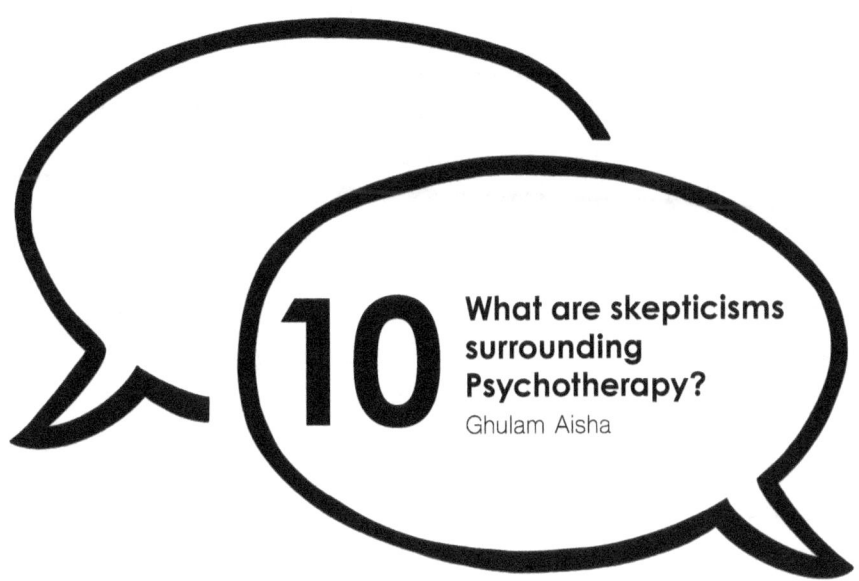

10 What are skepticisms surrounding Psychotherapy?

Ghulam Aisha

Abstract

Psychotherapy is extensively used in the medical world in order to treat a multitude of conditions. However, ever since the emergence and dominance of evidence-based treatments, the approach to psychotherapy has come into question. More specifically, a question whether psychotherapy is more art or science. Some argue that it needs to be an evidence based field of study while others argue that by doing so, we would be undermining the very thing that makes psychotherapy work. In addition, another skepticism surrounding psychotherapy is its effectiveness when treating children and adolescents. As a result of the different power dynamic between the provider and the patient and the heavy involvement of parents, children and young adolescents may not fully benefit from psychotherapeutic treatment as an adult would.

Introduction

As with any branch of psychology, there is a certain level of skepticism that surrounds it. The big question surrounding psychotherapy is whether it is more art than science. One the one hand, experts argue that therapists need to provide scientific studies to back what they do. On the other hand, it is argued that pushing for evidence undermines the very thing that makes psychotherapy work (Cary, 2004). This debate has significant implications because it determines how therapists are trained, techniques therapists can use, the kinds of therapy health insurers cover (Cary, 2004). Psychology is not a simple black and white field, rather a lot of confounding variables that have to be taken into consideration before a definitive diagnosis can be given. This paper will discuss the implications of adopting an evidence-based approach to psychotherapeutic treatments versus maintaining the status quo and will look at the problems associated with using psychotherapy to treat adolescents and children which renders them ineffective in that group.

BACKGROUND

Psychotherapy is used to treat psychiatric disorders in children. However, the response rate varies in that a significant proportion of patients may not experience a significant decrease in symptoms (Hong, 2015). Debates on psychotherapy and medicine began to emerge as more psychotropic medications were being used in clinical practice (Hong, 2015). An evidence-based approach entails that an instruction manual would be provided to therapists which consists of a compilation of short-term therapies that studies have shown work for a variety of mental disorders (Cary, 2004). For example, a standardized manual would quickly help people with panic attacks and phobias among other conditions (Cary, 2004). This is quickly becoming a common way for those working in health insurance. One of the leading health insurance companies in the United States already bases the coverage of psychotherapy based on the practice which therapists use and are backed by research (Cary, 2004). Other companies limit the number of sessions they cover for a certain diagnosis/condition based on what research says (Cary, 2004). Insurers may also require that therapists and patients both document the progress being made at sessions to provide evidence that therapy is working (Cary, 2004). This ensures that therapists are held accountable for the treatment they are providing. However, opposing parties believe that if therapists are required to provide proof to healthcare insurers, it breaches doctor-patient confidentiality. In addition, the treatment is far too complex to follow a standardized treatment approach. Forcing psychotherapy to follow a standardized format could result in people being denied the treatment they most need/require (Cary, 2004). The effectiveness of treatments was established in the past based on patient testimonies (Cary, 2004). However, when science began evolving and new drugs were introduced to help manage conditions, therapists were having to justify their methods with evidence. An empirical approach to psychotherapy brings with it accountability which increases the credibility of the field (Carry, 2004). This provides health care insurers and policy makers with the confidence that psychotherapy is a legitimate treatment that works and is not open-ended and subjective (Carry, 2004).

PSYCHOTHERAPY AND SCIENCE
There has been growing research into human physiology which increased our understanding of the "role of neurotransmitters, brain circuitry, and other neurobiological mechanisms to establish an empirical basis for psychiatric treatment" (Hong, p.158 , 2015). Due to an increased understanding of the cognitive and behavioral functions of the human body, there has been growing interest in the 'biological' basis of psychotherapy treatments. Behaviour and biology are intertwined and work best when used together. For example, a woman comes in with a condition resulting from amygdala degeneration which limits her ability to correctly identify fearful faces (Hong, 2015). This can be corrected by telling her to focus on the eye region of the face (Hong, 2015). This highlights the importance and interconnectivity of behaviour and biology and how they can be used together to treat conditions. Looking at the biological basis of psychotherapy, psychopharmacology and neurostimulation helps bridge the gap of treatment non-response providing relief to patients (Hong, 2015). Psychotherapeutic treatment plans that incorporate a biological basis are often found to be successful in treating conditions (Hong, 2015).

Psychotherapy is seen as being less scientific in comparison to traditional "hard" sciences like physics and chemistry (Holmes & Beins, 2009). The scientific nature of the field comes into question for a number of reasons. Hard sciences typically have objective measures such as volts or chemical levels (Lilienfeld and Gurung, 2012). Whereas, psychology depends on perception which is subjective. In addition, hard sciences typically have more well constructed and characterized research designs (Lilienfeld and Gurung, 2012). Lastly, it is difficult to duplicate results from a psychology study in comparison to studies from hard sciences (Lilienfeld and Gurung, 2012).

In order for the optimal implementation of evidence-based psychotherapy various factors need to be taken into account, such as relationships, providers, context and flexibility is required. Firstly, evidence-based psychotherapy is not useful without first considering the therapeutic relationship (Cook, Schwartz and Kaslow, 2017). A treatment method alone is ineffective if the therapeutic relationship is not a positive experience for a patient. Positive therapeutic relationships can be formed by using empathy, displaying cohesion among patients in a group therapy setting, forming a positive relationship in different settings such as group, youth and family settings, displaying collaboration, support and consensus when dealing with clients (Cook, Schwartz and Kaslow, 2017). Another important factor to consider is that of providers. When thinking about psychotherapy and providers the first ones to come to mind are psychologists. However, there are a number of other providers that also play a role in providing treatment to patients. These include but are not limited to nurses, professional counselors, physicians, social workers and graduate students (Cook, Schwartz and Kaslow, 2017). Given the number of people that work to facilitate the successful treatment of patients, there is a need to properly train them and provide ongoing supervision (Cook, Schwartz and Kaslow, 2017). It is important that healthcare providers are properly trained and possess the necessary experience and skills needed to assess patients and provide treatment.

Another important factor to consider is the context in which the implementation of evidence-based psychotherapy occurs. It can occur in a multitude of settings such as private practices, counseling centers, educational systems and Veterans Health Administration facilities (Cook, Schwartz and Kaslow, 2017). The problem is that a lot of psychotherapeutic treatments are designed for a single-diagnosis condition, when in fact a lot of patients come with requiring treating for a combination of conditions (Cook, Schwartz and Kaslow, 2017). This leads to the final factor which needs to be taken into account. Flexibility is a big part of a successful evidence-based psychotherapy practice. Flexibility can be used to build rapport with your patients, provide them with more autonomy to choose what kind of treatment they want and provide patients with the time they need to fully adopt the treatment plan, especially if they have a difficult time following multiple steps (Cook, Schwartz and Kaslow, 2017). In addition, it allows for the modification of a treatment plan to adjust for unexpected life circumstances that may occur (Cook, Schwartz and Kaslow, 2017).

CHILDREN AND ADOLESCENTS

There is skepticism around the benefits that psychotherapy offers children and adolescents. Some considerations that need to be taken into account when treating children and adolescents include the potential harmfulness of treatments, the fact that treatment of children typically involves parents and the role of developmental changes in the occurrence or recovery from emotional disturbance. Firstly, some treatments can potentially result in harmful effects. It is speculated that children with adverse childhood experiences (ACEs) may react poorly to basic treatments (Hupp et al., 2019). For example, forced holding in holding therapy. Forced holding is when a therapist or parents forcefully holds a child until they stop resisting, a fixed amount of time has elapsed or until the child makes eye contact (Hunt, n.d.). Without factoring in previous experiences, some treatments may pose more harm than good. In addition, children and adolescents are more vulnerable due to the problem of informed consent. Children and some adolescents may not be able to fully understand the benefits and risks associated with a treatment in order to provide informed consent (Hupp et al., 2019). Oftentimes when children do refuse treatment it may be interpreted as a symptom of an emotional problem such as oppositional behaviour (Hupp et al., 2019). Due to the limited authority that children and adolescents possess, they may be in a difficult position where they cannot seek help or report harmful events that take place in residential treatment centers.

Secondly, the involvement of parents in the treatment plan presents a problem. Alongside specific treatment for the child or adolescent experiencing the problem, parents are often trained to treat their child's emotional disturbances (Hupp et al., 2019). Emotional disturbances refer to a condition wherein one or more of the following characteristics is demonstrated over a long period of time to the point where it adversely impacts a child's educational performance. These include, "(A) An inability to learn that cannot be explained by intellectual, sensory, or health factors. (B) An inability to build or maintain satisfactory interpersonal relationships with peers and teachers. (C) Inappropriate types of behavior or feelings under normal circumstances. (D) A general pervasive mood of unhappiness or depression. (E) A tendency to develop physical symptoms or fears associated with personal or school problems" (Meier, n.d.). Some parents may be willing and happy to learn effective science-based strategies, however, others may resort to implausible, ineffective and pseudoscientific approaches when treating their children (Hupp et al., 2019). Parents possess great power over their children and are equally responsible for them. This may result in some parents overstepping and using potentially harmful treatments.

Lastly, another problem when treating children and adolescents is that of developmental change. As children grow they go through developmental changes that are most prominent during infancy, early childhood and then around the time of puberty (Hupp et al., 2019). As a result, therapeutic treatments need to be evaluated on an on-going basis while keeping patterns of developmental change in mind. For example, when treating autism spectrum disorder, you simply cannot assess the child on two different occasions throughout the year (Hupp et al., 2019). The child needs to be evaluated on an on-going basis due to developmental changes otherwise they may be getting treatment that is ineffective (Hupp et al., 2019). Pseudoscientific therapies do not consider developmental changes that occur and provide treatment according to that. Patients respond to experiences differently during different times in their lives. Something that worked when a patient was an infant will not work when they are a school-aged child. However, pseudoscientific therapies are based on recapitulation of development events (Hupp et al., 2019). This refers to the idea of having an older child repeat parts of their life as an infant, the child can go back and negotiate an earlier part of their development (Hupp et al., 2019). This most often occurs during treatments of emotional attachment.

Conclusion

Psychotherapy is not a new field of study but it is constantly evolving. However, the primary question of whether it is more of a science than an art has ignited a debate in the academic community. There are significant implications for this debate. Namely, it determines what treatments health insurance providers are willing to cover and holds providers accountable for the treatment they are giving. However, this may result in access to treatments being limited because often conditions are too complex to follow a standardized format. Another issue raised when looking at psychotherapy is the effectiveness it provides to children and adolescents. Oftentimes treatment for children and adolescents involves parents which results in a problem with them being authority figures and legitimate concerns from the patient not being taken seriously and instead being attributed to age. In addition, children and adolescents go through significant developmental changes and psychotherapeutic treatments need to be adjusted on an on-going basis in order to account for that. Otherwise, patients may be forced to continue with a treatment that is ineffective at helping them deal with the problem they have.

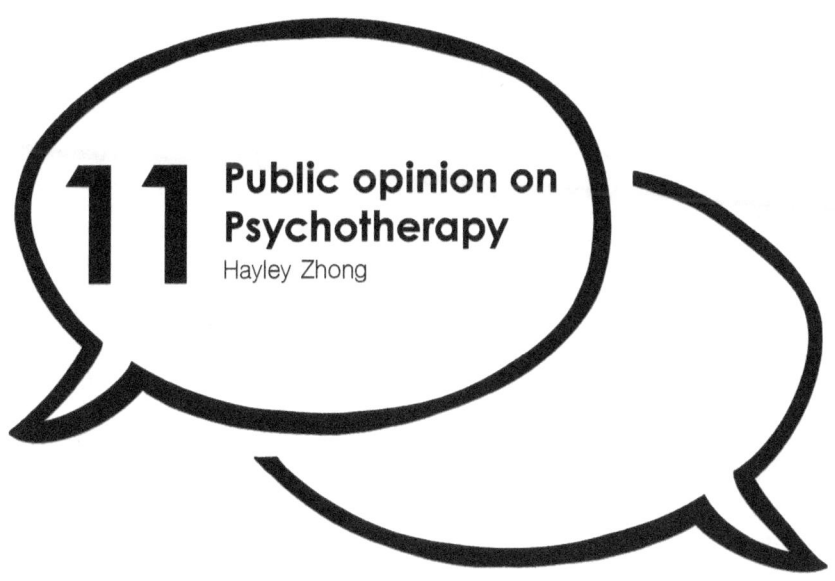

11 Public opinion on Psychotherapy
Hayley Zhong

We have come a long way from the days when mental asylums and trephining were used to "cure" the mentally ill. Despite the fact that many people believe in the efficacy of psychotherapy, most people nevertheless use drugs to manage their mental health and consider psychotherapy as a treatment choice primarily for those with diagnosable conditions. As agents of socialization, family, colleagues, culture, and the media have all played a part in the discussion of psychotherapy, casting either a positive or negative picture of the practice depending on the situation. Despite the fact that the dialogue around mental well being is becoming more open, individuals and the general public also have a stigma about psychotherapy that they have internalized by their agents of socializations, which contributes to both public and self stigma. This stigma surrounding psychotherapy plays a significant role in deterring people from seeking care and, as a result, affects the public's view of psychotherapy. Overall, many people could benefit from psychotherapy to help them navigate life, and daily exposure to practices is necessary to inform the public about the procedure and make it less taboo.

Attitudes toward mental health treatments have shifted dramatically over the years. For much of history, demonic possession was blamed for mental illness, leading to inadequate care and animosity toward those who needed help (Szasz, 1960). Due to this belief, treatments such as exorcism, in which a priest performs a ritual speaking prayers over the "possessed" being to rid a person, place or thing of the demonic spirits (Layton, 2020). Trephining, a procedure that involved drilling a small hold in a person's skull, was typical in more severe cases. It was thought that doing so would allow spirits to leave one's body (Lumen Learning, n.d.). Asylums became the medication of choice for those suspected of displaying "abnormal" behaviour in the 1700s. Instead of finding solutions, asylums were used to separate and shun persons by housing them and isolating them from the general public (Lumen Learning, n.d.). The living conditions in these places were inhumane, Patients would be bound by chains in their beds and isolated in windowless dungeons with no human interaction or contact. Antipsychotic drugs were not widely used until the 1960s.

Psychotherapy is usually well received by the general public but the number of people seeking this form of help is surprising when compared to its public reception. In an interview with 12 undergraduate students who had never had experienced psychotherapy, it was discovered that the benefits of psychotherapy, such as personal healing and development, were widely known and agreed upon (Hill et al., 2012). There was a general sense of dread in this audience about the prospect of being diagnosed. Despite the obstacles they recognized, such as self stigma, costs, and pressures to confront problems on their own, the general consensus was that if it was "necessary," it would be beneficial, and they would go (Hill et al., 2012). Following further inquiry, participants usually suggested that by "necessary," they were referring to diagnosable conditions that they believe to be "severe," such as depression and eating disorders were the ones that would necessitate psychotherapy (Hill et al., 2012). Many people did not think their problems or circumstances were severe enough to include psychotherapy (Snyder, Hill, & Derksen, 1972).

While the general public views psychotherapy as a successful way of dealing with issues, those who use it face a negative stigma (Ben-Porath,

2002). It is often regarded as a "last resort" for people after other approaches, such as talking to people in their lives show little to no success (Hinson & Swanson, 1993). This makes it difficult for people to pursue psychotherapy because of the public stigma that those who seek psychological treatment in this way are vulnerable or inferior to others (Dean & Chamberlain, 1993). Stigma is often cited as the most significant impediment to pursuing psychotherapy (Corrigan, 2004). People's pursuit for psychotherapy is hindered largely because of public stigma, which is compounded by self stigma, where the public stigma is internalized.

Critics see psychotherapy as an industry that profits from people's low self esteem, leading to the stigmatization of the profession (Angermeyer, 2017). Overall, people's confidence in psychotherapy and medicine has grown over the last twenty five years, as has the number of people seeking treatment from those options (Angermeyer, 2017). When opposed to prescription choices, psychotherapy is often underestimated compared to general practitioners offering pharmaceutical solutions.

A positive correlation exists between people's attitudes towards psychotherapy, their likelihood to reach out and receive professional help (Leech, 2007; Shaffer, Vogel, & Wei, 2006). People were also much more likely to participate in psychotherapy after being diagnosed with a diagnosable psychiatric condition and it also made them have a more optimistic outlook on the treatment's efficacy (Jorn, 2005). In general, the move toward greater acceptance of psychotherapy by the public is at odds with what is happening in clinical practice. Pharmacotherapy has been the most common treatment for many people in many countries (Angermeyer, 2017). The number of people undergoing psychotherapy in the United States of America decreased by 34% from 1998 to 2007. Medication for the same conditions, on the other hand, has risen by 23% (Gaudiano, 2013). It can be seen in over 33 studies that people's preference of psychotherapy over medication is three times greater (Gaudiano, 2013), so what is causing this trend of choosing medicinal solutions?

Psychotherapy has an image problem. Many are unaware and misinformed about the high level studies and research that supports the effectiveness of psychotherapy on those who take part. Studies have shown that the results of therapies and medications are comparable in the short term and that psychotherapy can be better in the long-term in the lives of patients and their family members and prevent relapses more effectively than if they were to be mediated. The lack of awareness about these psychotherapy studies from physicians, the insurance industry, those in government office, and the general public has fed into the assumption that medication and biologically based practices fed to us by the pharmaceutical industries are superior to psychologically based practices (Gaudiano, 2013). Many factors have contributed to the misrepresentation of psychotherapy in the minds of the general public including people in one's close circle, the media, personal experiences and interests, as well as cultural differences.

As key agents of socialization, family and friends have a great influence on one's beliefs and attitudes. Families are small, intimate groups that come into face to face contact frequently. This closeness and frequent exposure allows members to transfer beliefs, values, and attitudes to other members, including the meaning attached to their social surroundings, including that towards psychotherapy (Brym, 2019). These experiences vary between families and friends. Some exposures lead to a positive association with therapy, like witnessing the success of psychotherapy in a loved ones life after experiencing trauma, others may be negative. Seemingly insignificant moments like not having a school guidance counsellor may also lead to a negative perspective on psychotherapy due to lack of exposure of having a professional listen and help (Hill et al., 2012). If those who have a negative association with psychotherapy choose to partake with this attitude, a self fulfilling prophecy may occur, leading to a less effective treatment (Brym, 2019).

One's personal experiences also influenced their feelings towards psychotherapy. Given the general understanding of the role of a therapist to listen, support and give advice to the client and the client to be open to discussion and be willing to talk in a meaningful manner, many associated behaviours they exhibited on a day to day basis with that of a psychotherapy professional (Hill et al., 2012). Instances included being the mediator in their personal friend group, or having friends approach them for advice with their problems and feeling better after long discussion and having someone to listen and vocalize feelings to (Hill et al., 2012). This created a positive imagery of psychotherapy after first hand seeing the positive results after acting in a similar role as a psychotherapist.

Across cultures, there are variances about the opinion of psychotherapy. In some cultures, it can be seen as a sign of weakness. This stems from the beliefs that many cultures have of keeping information about one's personal life to oneself instead of exposing your troubles. Given this, talking to a therapist seems to violate this belief, making many less willing to take that step towards getting help and attempting to manage issues by themselves or only with close family and friends (Beker, 2015). In other cultures however, the idea of a therapist may not be taboo at all, and as normal as having a personal trainer. This may be due to the belief that having a neutral party listen about personal issues is expected (Beker, 2015).

Especially in the twenty-first century, the media has played a great role in public perception due to how popularized it is, leading to increased and repetitive exposure in the lives of the general public. It has a great influence on people's perception of what is reality. This can be a double edged sword because the portrayal of psychotherapy can either be accurate and progressive or inaccurate and create a spread of misinformation, contributing to the already prevalent stigma around psychotherapy. Doctor Mariel Buque from Psychology Today believes that positive, accurate depictions of therapy are greatly beneficial because it allows for regular people to get the opportunity to witness what therapy is like and its impact on a patient's life. Depictions like this allow for common misconceptions to be corrected and open up the discussion about mental health and seeking out psychotherapy as a means of help (Ryan, Ryan, & Ryan Rosenbaum, 2017). Mental health awareness campaigns like Bell Let's Talk Day and others run by organizations and governments have been increasing in numbers over the years, with the goal of educating the public and encouraging treatment sooner (Nemec, 2005).

George Gerbner's 1969 cultivation hypothesis states that there exists a positive correlation between the amount of exposure about a topic in the media and the acceptance of that particular topic, whether it be norms, values, attitudes, thoughts or beliefs (Potter & Chang, 1990). It believes that what is seen frequently in mass media, falls in line with the beliefs and values of people. In regards to the acceptance and image problem of psychotherapy, this is problematic because people are gathering information and developing their knowledge of mental health based on media sources that often depict psychotherapy practices in a negative light (Philo, 1994). These portrayals are often used for dramatic effect. It is not uncommon to see plots where a therapist does not engage in proper ethical practices like in CBC's Workin' Moms where Brad Heshinton, a psychiatrist, uses hypnotherapy to hurt and coerce women without their consent. Other times, it is the patients who are viewed as "insane" like in the show like in Netflix's Behind Her Eyes where it can be seen that past patients who sought after psychotherapy, acting destructively outside of their practice after receiving treatment. After seeing these interpretations of psychotherapists and those who are patients of psychotherapy time and time again, many internalize this message about the practice of psychotherapy itself, contributing to the negative stigma, and hindering people's willingness to seek help in this way (Hill et al., 2012). Contrastingly, large pharmaceutical companies pushing for medicinal solutions have the budget and marketing resources to push ads promoting their options, making people more likely to choose medicinal therapies rather than psychotherapy due to the large amount of exposure (Gaudiano, 2013).

Overall, these influences have contributed to a vast list of misconceptions about those who seek psychotherapy. This list about patients includes, but is not limited to: they are weak, crazy, don't have a solid support system, have been through a traumatic event, are in a bad mental state and have trouble letting things go (Kohli, 2016). These misconceptions are far from the truth and need to be broken down to stomp out the stigma. Psychotherapy can be valuable for anyone to live a better life, even those without any sort of diagnosis. The fact that many public health systems and insurance policies do not cover the cost of psychotherapy and only cover non-pharmaceutical options only feed into the negative narrative and inhibit people from seeking this form of help. This is carried on into the workforce where "sick days" are usually only able to be used for physical sickness. These negative perceptions aid in the stigma that is sopping three fourths of the estimated 50 million Americans living with mental illness from seeking psychological treatment (Center for Mental Health Services, 2000), the 98% of those who did not meet the diagnosable criteria with problems that do not turn to psychotherapy and the 44 percent of respondents in an Australian study claimed that they would feel a sense of embarrassment if they were to see a professional and take part in psychotherapy (Andrews, Issakidis, & Carter, 2001).

Popular culture has played a role in increasing awareness of mental health professionals, encouraging many to seek help. This has been creating more of a normalization of psychotherapy for individuals, even those without diagnosed conditions or those who do not seem to "warrant" psychotherapy to help. The age of those attending therapy has steadily decreased, with younger patients. This is occurring because what was once seen as a "last resort" is becoming more normalized, making parents bring their children into participate in psychotherapy as a positive coping mechanism to learn how to navigate day to day life and the conflict that comes along with it (Ryan, Ryan, & Ryan Rosenbaum, 2017).

Many who can reap benefits from psychotherapy are not seeking it due to the stigma associated. The way to improve public opinion of psychotherapy and encourage people to take action in seeking this form of help is to improve the public image of psychotherapy and create a better understanding of it. This includes stressing about how listening and understanding is at the core of this treatment, and how beneficial it is for everyone and that your relationship with your psychotherapy professional is not judgemental. This process includes opening up the conversations and normalizing confronting problems we face with a professional. In doing so, it would break down many barriers cited for not seeking help including feeling of shame, difficulty disclosing personal information with therapists and negative self perceptions (Deane & Todd, 1996). Following a shift in public opinion, we need institutions to follow to make it more accessible for everyone, which will again, increase the exposure of psychotherapy to individuals and lead to a happier, healthier population.

12 Future perspectives of Psychotherapy

Juwairia Razvi

Abstract

In this chapter, we will explore the importance of psychotherapy in the context of COVID-19 and the benefits of online treatments and interventions in reaching a wider population, including vulnerable populations, cutting on costs, and providing care and treatment in a more convenient and efficient way. Additionally, we will examine the future prospects of digital psychotherapy through development of AI based agents, robot therapy, its current uses and what researchers propose. We will explore the concerns with digital physiotherapy and algorithmic technology in the field of mental health as well as ethical concerns that are presented by experts.

Introduction

The American psychological association defines psychotherapy as a "collaborative treatment between an individual and a psychologist" (apa.org). It is also referred to as talk therapy (psychiatry.org) and is a safe and non-judgmental space for people to talk about their concerns, difficulties in their daily life, or personal thoughts and feelings (apa.org). A person may take about their problems and work together with the psychologist to identify and change troubling thoughts and behavior patterns allowing a person to feel better and increase mental and overall well-being (psychiatry.org & apa.org). There are various types of psychotherapy, and while they may be conducted slightly differently from each other, they all carry similar goals: helping individuals eliminate disruptive or negative thoughts and behaviours and replacing them with healthy behaviours, allowing a person to better manage their emotions, feelings, and find constructive ways of dealing with difficulties. Examples include CBT, which allows a person to replace harmful thoughts with those that are useful and healthy, helping with treating depression, anxiety, and eating disorders. Interpersonal therapy can help people learn healthy ways to express and better communicate emotions, helping with grief and depression (psychiatry.org).

Mental health and COVID-19

The COVID-19 pandemic has opened doors to many challenges among frontline workers and the general population such as deteriorating mental health and difficulties in managing existing mental health concerns. Changes in both personal and social life including job loss for many, insecurity, social isolation, school closures, combined with inadequate resources and government response to growing mental health concerns have translated into emotional outcomes such as heightened anxiety, depression, fear, frustration, psychiatric conditions, and unhealthy behaviors including excessive substance use (Pfefferbaum & North, 2020). The delivery of psychosocial services such as counselling and therapy has also been greatly affected by stay-at-home orders and quarantine. Services that were once delivered in primary care settings now have to be offered virtually, and this presents additional difficulties to both the mental health professional and the patient that they must navigate (Pfefferbaum & North, 2020). While those experiencing milder psychosocial outcomes are advised to use supportive interventions to cope with mental health challenges such self-care, scheduling, taking breaks and practicing meditation through the use of easily accessible apps and online resources, other individuals need more specialized care such as referral to formal mental health evaluation and care (Pfefferbaum & North, 2020).

Various studies have assessed the impacts on mental health of those who are most vulnerable amidst the pandemic: frontline workers, single parents, students especially international students, low income families, and many other groups. Many healthcare providers experience emotional distress in the current pandemic due to concerns about getting infected 'given their risk of exposure to the virus', concern about possibly infecting loved ones, shortages of resources, longer work hours, and involvement in difficult (social and ethical) decisions (Pfefferbaum & North, 2020). It is important that support for healthcare workers be integrated into their daily routines and in their workplaces to ensure that their mental and physical health is in check and they are capable of taking care of sick patients.

Due to COVID-19 restrictions, individuals have less access to in-person and direct psychosocial services. Accessing healthcare services is limited in many places due to restrictions through a general practitioner, family doctor, or in a hospital setting, therefore Pfefferbaum & North (2020) advice that it is important that healthcare staff in clinic and hospital settings be provided mental health training so they may provide basic assessment of psychosocial concerns and appropriate referrals. "Monitoring psychosocial needs, delivering support to their patients, health care providers, and the [general] public should be integrated into general pandemic health care plans" on regional, national and international levels (Pfefferbaum & North, 2020). Ensuring quality support for healthcare providers can allow them to maintain healthy lifestyles, behaviours and thoughts amid the chaos of their daily lives, so that they may work efficiently and provide optimal care to COVID-19 and other patients.

The COVID-19 pandemic has also sparked conversations around the use of videoconferencing and online psychotherapy, with arguments on either end. McDonald et al. (2019) discuss evidence in support of online based CBT such as more accessibility. Lower income countries or regions may have fewer access to resources for in-person therapy including lack of specialized staff and may face financial restraints due to high cost of services. Additionally, McDonald et al. (2019) emphasize that the arbitrary threshold or standard for accessing need for psychotherapy may not be accurate for certain social groups, showing that they may not be in need of services, while in reality, these individuals would be highly encouraged to seek out treatment. Additionally, online psychotherapy is beneficial for individuals with physical constraints such as physical disability, imprisonment or social anxiety which inhibit their ability to attend multiple therapy sessions or be thoroughly involved (McDonald et al., 2019). Online modules, often unguided therapy is a great starting point for psychotherapy care for different groups as discussed previously, reducing high therapist costs. It may be followed by personalized support digitally through text messages, phone calls or video conferencing, or in certain cases, personal contact with the therapist. A large number of people such as vulnerable groups, individuals living sparsely in large geographical areas or hard to reach areas can be easily reached, saving on therapist costs, patient time, and essential resources. Digital therapy is an efficient and promising solution to improve mental health during the pandemic by reaching millions of people worldwide through use of online forums and apps. Current and future applications of online psychotherapy programs and apps will be discussed further in the next section.

Mental health applications

Pre-pandemic, there have been various successful interventions and developments in the field of digital psychotherapy, from supporting students, new mothers, multi-cultural groups, Indiegnous groups and other groups that face unique concerns. About 15% of women experience postpartum mood disorders, (Yang et al., 2019) with barriers to access physical therapy which include childcare responsibilities, physical inability, or cost of travel and treatment. These barriers may lead to perinatal mental illness and worsening health of both mother and child(ren) (Yang et al., 2019). In a randomized control trial, Yang et al. (2019) examines video conferencing-based therapy in comparison to the typical office-based therapy. It was found that there were high levels of attendance and satisfaction with videoconferencing based therapy. Although there was a limitation regarding a small sample size, Yang et al. (2019) emphasize the need for a larger study to better understand the relationship between videoconferencing therapy and overall attendance in comparison to office-based treatment combined with videoconferencing therapy and overall attendance. This will allow researchers and experts to study the effects of videoconferencing as a feasible option for psychotherapy among women in the postpartum period.

Additionally, a study by Huberty et al. (2019) examined the effectiveness of the "Calm" app in reducing stress and anxiety levels among university students. The app involves procedures based in mindfulness-based stress reduction and mindfulness-based cognitive therapy, both of which are effective ways to reduce overall stress while increasing self-compassion, mindfulness, and overall mental wellbeing (Huberty et al., 2019). Traditionally, these practices are time-consuming, and the goal of mental health experts and app developers is to curate treatments and practices that are quick and accessible for students through mobile mental health applications (Huberty et al., 2019) The study notes significant improvements in self-compassion, reduced stress levels, and mindfulness among students that used Calm versus students who did not use the app. Huberty et al. (2019) also predict promising results in improvement of sleep quality due to students' reduced stress, and daily meditation implementing healthy behaviours. Students were also satisfied with using the different components of Calm and found it helpful in their daily routine, mainly due to aspects such as accountability, tracking progress, reminders, notifications, and overall convenience (Huberty et al., 2019). Similarly apps such as Togatherall (formally known as Big White Wall), Headspace and Happify are all useful apps for students to manage stress, practice mindfulness, learn healthy habits and behaviours, and receive counselling. Many universities provide students with free access to some of these apps and other mental health online sessions.

Digital psychotherapy is promising exciting development through AI, user interface and robot-led treatment (Fiske et al., 2019). These virtual therapists "include an artificially intelligent algorithm that responds independently of any expert human guidance to the patient" (Fiske et al., 2019). They are different from online psychotherapy and health-based apps which include some aspect of human interaction with a therapist (Fiske et al., 2019). Currently, AI psychotherapeutic agents such as Tess and Woebot are being developed to help patients navigate depression and anxiety (Fiske et al., 2019). Woebot is able to interact with the patient as a virtual psychotherapist, discussing and working with them to recognize emotions, thoughts and to develop healthy skills while Tess can recognize expressions that indicate emotional distress and help the patient work through them (Fiske et al., 2019). "companion bots, such as Paro can act as at-home health care assistants by responding to speech and movement and can hold a conversation, providing essential help to elderly individuals or clients dealing with depressive symptoms (Fiske et al., 2019). Studies show that these interactions with these robots helps reduce stress, loneliness and improve mood and overall happiness (Fiske et al., 2019). AI robots also provide opportunities for different forms of engagement with children suffering from autism spectrum disorders, forming a social bond with children, integrating therapy interventions, and helping children improve social skills (Fiske et al., 2019). These robots are still being developed and research is still in its preliminary settings. It would take many years for large-scale research, formal consensus regarding ethical use as well as discussions regarding funding and accessibility to form public policy and wider therapeutic use of AI agents and assistive robots.

CONCERNS WITH DEVELOPMENTS IN DIGITAL PSYCHOTHERAPY
Through analysis of proposed interventions can lead to early identification of ethical issues and can help researchers and AI developers "consider these concerns in the design and construction of the next generation of AI agents and robots for mental health" (Fiske et al., 2019). Constructive and in-depth analysis will ensure that we will responsibly use these applications in clinical settings and to advance mental health on a population-based level. AI agents, chatbots and assistive robots will be closely monitoring mood, feelings, verbal and physical cues of clients and its necessary that clear standards around confidentiality, information privacy, and data security regarding the use of data that is collected be formulated and made public (Fiske et al., 2019). At the moment, "An Ethical Framework for a Good AI Society: Opportunities,

Risks, Principles, and Recommendations" sets guidelines for the use of AI technology in public, but there is no internationally recognized thorough set of guidelines that establish recommendations on ethical use regarding, or even structural guidance on the development of digital mental health services (Fiske et al., 2019). AI and robot-based mental health services are not a substitute for established and rigorous mental health care available in our healthcare systems (Fiske et al., 2019). Care should be taken as to not replace existing mental health services such as psychotherapy with digital forms of psychotherapy; instead, researchers must incorporate it slowly, accessing its benefits and drawbacks frequently (Fiske et al., 2019).

While digital psychotherapy presents many advantages, Essig (2019) argues on " remove the human from the psychotherapy equation". He believes that incorporating algorithmic technology and advances in digital or online psychotherapy attracts entrepreneurs and experts to chase after high profits, rather than to "enrich therapeutic relationships" and improve quality of care for growing mental health concerns among the population (Essig, 2019). He argues that another major limitation of digital psychotherapy is "instruction-manual therapies", or therapies designed as a set of procedures and stages that do not reflect real-world, multi-dimensional issues complex that people face as well as the unique and personalized treatment plans formed in partnership with the therapist. Essig (2019) believes that researchers are going down a tricky path with evidence-based treatments or treatments that are grounded in research based on controlled trials instead of real-life conditions that present ineffective solutions. In recent decades, psychotherapy research has put a lot of attention to outcomes and less on the therapeutic processes, placing more emphasis on research through clinical trials (Goldfried in Aftab, 2020). Instead Goldfried believes that clinical observation is essential to more constructive treatments and overall development in the field of psychotherapy (Aftab, 2020). Insurance companies give preference to these " instruction-manual treatments" that target specific acute symptoms rather than the underlying cause of psychosocial issues and mental illnesses (Essig, 2019). This means more profits for companies and researchers, but it comes at the cost of poor and degreading mental health in our communities, especially our vulnerable communities (Essig, 2019).

As a rebuttal to an advantage presented earlier in the chapter, these technologies are not an 'end all be all' solution to lack of physical mental health resources in regions which have limited resources and funding. How can we be certain that technological corporations have the best interest of communities in mind and are thoroughly considering the ethical implications regarding affordability and equality in access, especially among marginalized and vulnerable individuals? Fiske et al. (2019) questions if an AI support agent does detect speech or feelings of a client that are concerning, can it provide additional support or will it refer the client to a therapist, and what happens if there are no therapists in the region. How can we continue to develop innovative technologies in the field of psychotherapy while continuing to pay attention to growing concerns of data misuse, security, and general inequalities in mental healthcare access among the population. Overall, establishing regulations around privacy, autonomy and data security within AI and robot-based mental health services is extremely important and it is important to question existing and developing technologies and set concrete guidelines for research and application.

References

CHAPTER ONE

Cohut, M. (2020, November 16). What is bloodletting, and why was it a popular therapy? Medical News Today. https://www.medicalnewstoday.com/articles/bloodletting-why-doctors-used-to-bleed-their-patients-for-health

Hayen, T. (2016). Ancient Egypt and Modern Psychotherapy: Sacred Science and the Search for Soul. In Ancient Egypt and Modern Psychotherapy: Sacred Science and the Search for Soul (p. 156). https://doi.org/10.4324/9781315650593

Ikiuga, M. N., & Ciaravino, E. A. (2007). Psychosocial conceptual practice models in occupational therapy: Building adaptive capability. Mosby.

Norcross, J. C., VandenBos, G. R., & Freedheim, D. K. (2011). History of psychotherapy: Continuity and change (2nd ed.) (2nd ed.). American Psychological Association. https://doi.org/10.1037/12353-000

CHAPTER TWO

Bradford, A. (2016, May 12). Sigmund Freud: Life, Work & Theories. LiveScience. https://www.livescience.com/54723-sigmund-freud-biography.html.

Public Broadcasting Service. (2002). Young Dr. Freud . PBS. https://www.pbs.org/youngdrfreud/.
BBC. (2014). History - Sigmund Freud. BBC. http://www.bbc.co.uk/history/historic_figures/freud_sigmund.shtml.

Cherry, K. (2020a, November 19). How Sigmund Freud Spent the Last Years of His Life. Verywell Mind. https://www.verywellmind.com/when-did-sigmund-freud-die-2795852.

Britannica, T. Editors of Encyclopaedia (2021, January 11). Josef Breuer. Encyclopedia Britannica. https://www.britannica.com/biography/Josef-Breuer

Cherry, K. (2020, April 24). Anna O. and Her Influence on Psychoanalysis. Verywell Mind. https://www.verywellmind.com/who-was-anna-o-2795857.

JUNG, C. (1989). Analytical Psychology: Notes of the Seminar Given in 1925 (McGUIRE W., Ed.). PRINCETON, NEW JERSEY: Princeton University Press. Retrieved May 15, 2021, from http://www.jstor.org/stable/j.ctt7rgpf

Launer, J. (2005). Anna O and the 'talking cure.' QJM: An International Journal of Medicine, 98(6), 465–466. https://doi.org/10.1093/qjmed/hci068

Waude, A. (2016, March 17). The Life And Case Study Of Anna O: How Sigmund Freud Was Influenced By One Woman's Expe. Psychologist World. https://www.psychologistworld.com/freud/anna-o-case-study-freud.

Jay, M. Evan (2021, May 2). Sigmund Freud. Encyclopedia Britannica. https://www.britannica.com/biography/Sigmund-Freud

Jones, J. (n.d.). Free Associations Method. http://www.freudfile.org/psychoanalysis/free_associations.html.

Marmor, J. (1970). Limitations of Free Association. Archives of General Psychiatry, 22(2), 160. https://doi.org/10.1001/archpsyc.1970.01740260064009

Schachter J. (2018) Free Association: From Freud to Current Use—The Effects of Training Analysis on the Use of Free Association, Psychoanalytic Inquiry, 38:6, 457-467, DOI: 10.1080/07351690.2018.1480231

Freud, Sigmund. (1896a). L'hérédité et l'étiologie des név-roses. Revue neurologique, 4: 161-169; GW, 1: 407-422; Heredity and the aetiology of the neuroses. SE, 3: 141-156.

Israëls, H., & Schatzman, M. (1993). The seduction theory. History of Psychiatry, 4(13), 23–59. https://doi.org/10.1177/0957154X9300401302

Burton, E. S. (2015). Sigmund Freud. Sigmund Freud | Institute of Psychoanalysis. https://psychoanalysis.org.uk/our-authors-and-theorists/sigmund-freud.

McLeod, S. A. (2008, October 24). Little hans - freud (1909). Simply Psychology. https://www.simplypsychology.org/little-hans.html

Thapaliya, S. (2017). The case of rat man: A psychoanalytic understanding of obsessive-compulsive disorder. Journal of Mental Health and Human Behavior , 22(2), 132–135. https://doi.org/10.4103/jmhhb.jmhhb_22_17

Billig, M. (1997). Freud and Dora. Theory, Culture & Society, 14(3), 29–55. https://doi.org/10.1177/026327697014003002

Nolan, M., & O'Mahony, K. (1987). Freud and Feminism. Studies: An Irish Quarterly Review, 76(302), 159-168. Retrieved May 15, 2021, from http://www.jstor.org/stable/30090854

Schafer, R. (1974). Problems in Freud's Psychology of Women. Journal of the American Psychoanalytic Association, 22(3), 459–485. https://doi.org/10.1177/000306517402200301

Ellis H. (1939). Freud's Influence on the Changed Attitude Toward Sex. American Journal of Sociology, 45(3), 309-317. Retrieved May 15, 2021, from http://www.jstor.org/stable/2769848

Murphy, G. (1956). The current impact of Freud upon psychology. American Psychologist, 11(12), 663–672. https://doi.org/10.1037/h0048812

Ruitenbeek, H. M. (1967). Heirs to Freud: essays in Freudian psychology. Grove Press, Inc.

CHAPTER THREE

American Psychological Association. (2017). What Is Psychotherapy? American Psychological Association. https://www.apa.org/ptsd-guideline/patients-and-families/psychotherapy#:~:text=Psychotherapy%20involves%20communication%20between%20patients,less%20anxious%2C%20fearful%20or%20depressed.

American Psychological Association. (n.d. a). Countertransference. In APA dictionary of psychology. Retrieved May 15, 2021, from https://dictionary.apa.org/countertransference

American Psychological Association. (n.d. b). Transference. In APA dictionary of psychology. Retrieved May 15, 2021, from https://dictionary.apa.org/transference

Bateman, A., Pedder, J., & Brown, D. (2000). Introduction to Psychotherapy: An Outline of Psychodynamic Principlesand Practice. Third Edition. Brunner Routledge.

Bruch, H. (1974). Learning psychotherapy rationale and ground rules. Harvard Univ. Press.

Cawley, R.H. (1977) 'The teaching of psychotherapy', Association of University Teachers of Psychiatry Newsletter January: 19–36.

Tsai, M., Yoo, D., Hardebeck, E. J., Loudon, M. P., & Kohlenberg, R. J. (2019). Creating safe, evocative, attuned, and mutually vulnerable therapeutic beginnings: Strategies from functional analytic psychotherapy. Psychotherapy, 56(1), 55-61

CHAPTER FOUR

Cognitive Behavioural Play Therapy. Retrieved 16 May 2021, from https://www.drfountain.ca/cognitive-behavioual-play-ther

Definition of Cognitive therapy. Retrieved 16 May 2021, from https://www.medicinenet.com/cognitive_therapy/definition.htm

Different approaches to psychotherapy. (2019). Retrieved 16 May 2021, from https://www.apa.org/topics/psychotherapy/approaches

Eelen, P. (2018). Behaviour Therapy and Behaviour Modification Background and Development. Psychologica Belgica, 58(1), 184. doi: 10.5334/pb.450

Gotter, A. (2018). Behavioral Therapy: Definition, Types, and effectiveness. Retrieved 16 May 2021, from https://www.healthline.com/health/behavioral-therapy#types

Psychoanalytic Therapy. (2021 - A). Retrieved 16 May 2021, from https://www.psychologytoday.com/ca/therapy-types/psychoanalytic-therapy#:~:text=Psychoanalytic%20therapy%20is%20a%20form,-to%20the%20surface%20and%20examined

Psychoanalytic Therapy. (2021 - B). Retrieved 16 May 2021, from https://www.counselling-directory.org.uk/psychoanalytical.html#thehistoryofpsychoanalysis

Rauch, J. (2016). Different Types of Therapy [Psychotherapy]: Which is Best For You? | Talkspace. Retrieved 16 May 2021, from https://www.talkspace.com/blog/different-types-therapy-psychotherapy-best/

Raypole, C. (2019). Humanistic Therapy: Definition, Examples, Uses, Finding a Therapist. Retrieved 16 May 2021, from https://www.healthline.com/health/humanistic-therapy

Shedler, J. (2010). The efficacy of psychodynamic psychotherapy. American Psychologist, 65(2), 98-109. doi: 10.1037/a0018378

Shiel Jr., W. (2021). Definition of Cognitive therapy. Retrieved 16 May 2021, from https://www.medicinenet.com/cognitive_therapy/definition.htm

CHAPTER FIVE

American Psychological Association. (n.d.). Behavior. APA Dictionary of Psychology. Retrieved May 16, 2021, from https://dictionary.apa.org/behavior

Ardito, R. B., & Rabellino, D. (2011). Therapeutic Alliance and Outcome of Psychotherapy: Historical Excursus, Measurements, and Prospects for Research. Frontiers in Psychology, 2. https://doi.org/10.3389/fpsyg.2011.00270

Center for Substance Abuse Treatment. (1999a). Chapter 6—Brief Humanistic and Existential Therapies. In Brief Interventions and Brief Therapies for Substance Abuse. Substance Abuse and Mental Health Services Administration (US). https://www.ncbi.nlm.nih.gov/books/NBK64939/

Center for Substance Abuse Treatment. (1999b). Chapter 7—Brief Psychodynamic Therapy. In Brief Interventions and Brief Therapies for Substance Abuse. Substance Abuse and Mental Health Services Administration (US). https://www.ncbi.nlm.nih.gov/books/NBK64952/

Chow, P. I., Wagner, J., Lüdtke, O., Trautwein, U., & Roberts, B. W. (2017). Therapy experience in naturalistic observational studies is associated with negative changes in personality. Journal of Research in Personality, 68, 88–95. https://doi.org/10.1016/j.jrp.2017.02.002

Connolly, M. B., Crits-Christoph, P., Shappell, S., Barber, J. P., & Luborsky, L. (1998). Therapist Interventions in Early Sessions of Brief Supportive-Expressive Psychotherapy for Depression. The Journal of Psychotherapy Practice and Research, 7(4), 290–300.

Doroshow, D. B. (2010). An Alarming Solution: Bedwetting, Medicine, and Behavioral Conditioning in Mid-Twentieth-Century America. Isis, 101(2), 312–337. https://doi.org/10.1086/653095

Dumper, K., Jenkins, W., Lacombe, A., Lovett, M., & Perimutter, M.

(2019). 7.1 What is Cognition? – Introductory Psychology. Washington State University. https://opentext.wsu.edu/psych105/chapter/7-2-what-is-cognition/

Eelen, P., Crombez, G., & Van den Bergh, O. (2018). Behaviour Therapy and Behaviour Modification Background and Development. Psychologica Belgica, 58(1), 184–195. https://doi.org/10.5334/pb.450

Elliott, R. (2002). The effectiveness of humanistic therapies: A meta-analysis. In Humanistic psychotherapies: Handbook of research and practice (pp. 57–81). American Psychological Association. https://doi.org/10.1037/10439-002

Gazzola, N. (Nick), & Stalikas, A. (2004). Therapist Interpretations and Client Processes in Three Therapeutic Modalities: Implications for Psychotherapy Integration. Journal of Psychotherapy Integration, 14, 397–418. https://doi.org/10.1037/1053-0479.14.4.397

Goodkind, M., Eickhoff, S. B., Oathes, D. J., Jiang, Y., Chang, A., Jones-Hagata, L. B., Ortega, B. N., Zaiko, Y. V., Roach, E. L., Korgaonkar, M. S., Grieve, S. M., Galatzer-Levy, I., Fox, P. T., & Etkin, A. (2015). Identification of a Common Neurobiological Substrate for Mental Illness. JAMA Psychiatry, 72(4), 305–315. https://doi.org/10.1001/jamapsychiatry.2014.2206

Kaczkurkin, A. N., & Foa, E. B. (2015). Cognitive-behavioral therapy for anxiety disorders: An update on the empirical evidence. Dialogues in Clinical Neuroscience, 17(3), 337–346.

Kandel, E. R. (1998). A New Intellectual Framework for Psychiatry. American Journal of Psychiatry, 155(4), 457–469. https://doi.org/10.1176/ajp.155.4.457

Malhotra, S., & Sahoo, S. (2017). Rebuilding the brain with psychotherapy. Indian Journal of Psychiatry, 59(4), 411–419. https://doi.org/10.4103/0019-5545.217299

Matusiewicz, A. K., Hopwood, C. J., Banducci, A. N., & Lejuez, C. W. (2010). The Effectiveness of Cognitive Behavioral Therapy for Personality Disorders. The Psychiatric Clinics of North America, 33(3), 657–685. https://doi.org/10.1016/j.psc.2010.04.007

Neuman, F. (2015, November 22). Standard Interpretations in Psychotherapy | Psychology Today Canada. Psychology Today. https://www.psychologytoday.com/ca/blog/fighting-fear/201511/standard-interpretations-in-psychotherapy

Picou, P., Adenuga, P., Ellison, K., & Davis, T. E. (2020). Specific Phobias in Children and Adolescents. In Reference Module in Neuroscience and Biobehavioral Psychology. Elsevier. https://doi.org/10.1016/B978-0-12-818697-8.00041-8

Pollack, J., Flegenheimer, W., & Winston, A. (1991). Brief Adaptive Psychotherapy. In Handbook of short-term dynamic psychotherapy (pp. 199–219). Basic Books.

Raypole, C., & Legg, T. J. (2019, March 1). A Guide to Different Types of Therapy. Healthline. https://www.healthline.com/health/types-of-therapy

Rehman, I., Mahabadi, N., Sanvictores, T., & Rehman, C. I. (2021). Classical Conditioning. In StatPearls. StatPearls Publishing. http://www.ncbi.nlm.nih.gov/books/NBK470326/

Sargın, A. E., Özdel, K., & Türkçapar, M. H. (2017). Cognitive-Behavioral Theory and Treatment of Antisocial Personality Disorder. In Psychopathy—New Updates on an Old Phenomenon. IntechOpen. https://doi.org/10.5772/intechopen.68986

Schut, A., Castonguay, L., Flanagan, K., Yamasaki, A., Barber, J., Bedics, J., & Smith, T. (2005). Therapist interpretation, patient–therapist interpersonal process, and outcome in psychodynamic psychotherapy for avoidant personality disorder. Psychotherapy: Theory, Research,

Practice, Training, 42, 494–511. https://doi.org/10.1037/0033-3204.42.4.494

Shafran, R., Bennett, S. D., & McKenzie Smith, M. (2017). Interventions to Support Integrated Psychological Care and Holistic Health Outcomes in Paediatrics. Healthcare, 5(3). https://doi.org/10.3390/healthcare5030044

Siegel, S. (2001). Pavlovian Conditioning and Drug Overdose: When Tolerance Fails. Addiction Research & Theory, 9(5), 503–513. https://doi.org/10.3109/16066350109141767

Wynn, G. H., & Ursano, R. J. (2017). Posttraumatic Stress Disorder. In Reference Module in Neuroscience and Biobehavioral Psychology. Elsevier. https://doi.org/10.1016/B978-0-12-809324-5.05378-5

CHAPTER SIX
Chakhssi, F., Zoet, J. M., Oostendorp, J. M., Noordzij, M. L., & Sommers-Spijkerman, M. (2021). Effect of psychotherapy for borderline personality disorder on quality of life: A systematic review and meta-analysis. Journal of Personality Disorders, 35(2), 255–269. https://doi.org/10.1521/pedi_2019_33_439

Crits-Christoph, P., Connolly Gibbons, M. B., Ring-Kurtz, S., Gallop, R., Stirman, S., Present, J., Temes, C., & Goldstein, L. (2008). Changes in Positive Quality of Life over the Course of Psychotherapy. Psychotherapy (Chicago, Ill.), 45(4), 419–430. https://doi.org/10.1037/a0014340

de la Rie, S. M., Noordenbos, G., & van Furth, E. F. (2005). Quality of life and eating disorders. Quality of Life Research: An International Journal of Quality of Life Aspects of Treatment, Care and Rehabilitation, 14(6), 1511–1522. https://doi.org/10.1007/s11136-005-0585-0

DeJong, H., Oldershaw, A., Sternheim, L., Samarawickrema, N., Kenyon, M. D., Broadbent, H., Lavender, A., Startup, H., Treasure, J., &

Schmidt, U. (2013). Quality of life in anorexia nervosa, bulimia nervosa and eating disorder not-otherwise-specified. Journal of Eating Disorders, 1(1), 43. https://doi.org/10.1186/2050-2974-1-43

Deter, H.-C. (2012). Psychosocial interventions for patients with chronic disease. BioPsychoSocial Medicine, 6(1), 2. https://doi.org/10.1186/1751-0759-6-2

Gil-González, I., Martín-Rodríguez, A., Conrad, R., & Pérez-San-Gregorio, M. Á. (2020). Quality of life in adults with multiple sclerosis: A systematic review. BMJ Open, 10(11), e041249. https://doi.org/10.1136/bmjopen-2020-041249

Grenon, R., Schwartze, D., Hammond, N., Ivanova, I., Mcquaid, N., Proulx, G., & Tasca, G. A. (2017). Group psychotherapy for eating disorders: A meta-analysis. The International Journal of Eating Disorders, 50(9), 997–1013. https://doi.org/10.1002/eat.22744

Haraldstad, K., Wahl, A., Andenæs, R., Andersen, J. R., Andersen, M. H., Beisland, E., Borge, C. R., Engebretsen, E., Eisemann, M., Halvorsrud, L., Hanssen, T. A., Haugstvedt, A., Haugland, T., Johansen, V. A., Larsen, M. H., Løvereide, L., Løyland, B., Kvarme, L. G., Moons, P., ... Helseth, S. (2019). A systematic review of quality of life research in medicine and health sciences. Quality of Life Research, 28(10), 2641–2650. https://doi.org/10.1007/s11136-019-02214-9

Health Quality Ontario. (2017). Psychotherapy for Major Depressive Disorder and Generalized Anxiety Disorder: A Health Technology Assessment. Ontario Health Technology Assessment Series, 17(15), 1–167.

IsHak, W. W., Ha, K., Kapitanski, N., Bagot, K., Fathy, H., Swanson, B., Vilhauer, J., Balayan, K., Bolotaulo, N. I., & Rapaport, M. H. (2011). The Impact of Psychotherapy, Pharmacotherapy, and Their Combination on Quality of Life in Depression. Harvard Review of Psychiatry, 19(6), 277–289. https://doi.org/10.3109/10673229.2011.630828

Jyrä, K., Knekt, P., & Lindfors, O. (2017). The impact of psychotherapy treatments of different length and type on health behaviour during a five-year follow-up. Psychotherapy Research, 27(4), 397–409. https://doi.org/10.1080/10503307.2015.1112928

Kolovos, S., Kleiboer, A., & Cuijpers, P. (2016). Effect of psychotherapy for depression on quality of life: Meta-analysis. The British Journal of Psychiatry, 209(6), 460–468. https://doi.org/10.1192/bjp.bp.115.175059

Kulacaoglu, F., & Kose, S. (2018). Borderline Personality Disorder (BPD): In the Midst of Vulnerability, Chaos, and Awe. Brain Sciences, 8(11). https://doi.org/10.3390/brainsci8110201

Laake-Geelen, C. C. M. van, Smeets, R. J. E. M., Quadflieg, S. P. A. B., Kleijnen, J., & Verbunt, J. A. (2019). The effect of exercise therapy combined with psychological therapy on physical activity and quality of life in patients with painful diabetic neuropathy: A systematic review. Scandinavian Journal of Pain, 19(3), 433–439. https://doi.org/10.1515/sjpain-2019-0001

Lim, J.-A., Choi, S.-H., Lee, W. J., Jang, J. H., Moon, J. Y., Kim, Y. C., & Kang, D.-H. (2018). Cognitive-behavioral therapy for patients with chronic pain. Medicine, 97(23). https://doi.org/10.1097/MD.0000000000010867

Linardon, J., & Brennan, L. (2017). The effects of cognitive-behavioral therapy for eating disorders on quality of life: A meta-analysis. The International Journal of Eating Disorders, 50(7), 715–730. https://doi.org/10.1002/eat.22719

Lorenz, T. K. (2021). Predictors and impact of psychotherapy side effects in young adults. Counselling and Psychotherapy Research, 21(1), 237–243. https://doi.org/10.1002/capr.12356

Marinelli, C., Savarino, E., Inferrera, M., Lorenzon, G., Rigo, A., Ghisa, M., Facchin, S., D'Incà, R., & Zingone, F. (2019). Factors Influencing Disability and Quality of Life during Treatment: A Cross-Sectional Study on IBD Patients. Gastroenterology Research and Practice, 2019, e5354320. https://doi.org/10.1155/2019/5354320

Sturgeon, J. A. (2014). Psychological therapies for the management of chronic pain. Psychology Research and Behavior Management, 7, 115–124. https://doi.org/10.2147/PRBM.S44762

White, C. A. (2001). Cognitive behavioral principles in managing chronic disease. Western Journal of Medicine, 175(5), 338–342.

CHAPTER SEVEN
Adler, N. E., Page, A. E., & Setting, I. of M. (US) C. on P. S. to C. P. in a C. (2008a). Psychosocial Health Services. In Cancer Care for the Whole Patient: Meeting Psychosocial Health Needs. National Academies Press (US). https://www.ncbi.nlm.nih.gov/books/NBK4017/

Adler, N. E., Page, A. E., & Setting, I. of M. (US) C. on P. S. to C. P. in a C. (2008b). The Psychosocial Needs of Cancer Patients. In Cancer Care for the Whole Patient: Meeting Psychosocial Health Needs. National Academies Press (US). https://www.ncbi.nlm.nih.gov/books/NBK4011/

American Psychiatric Association. (n.d.-a). What Is Depression? American Psychiatric Association. Retrieved May 16, 2021, from https://www.psychiatry.org/patients-families/depression/what-is-depression American Psychiatric Association. (n.d.-b). What Is PTSD? American Psychiatric Association. Retrieved May 14, 2021, from https://www.psychiatry.org/patients-families/ptsd/what-is-ptsd

Anxiety and Depression Association of America. (n.d.). Generalized Anxiety Disorder (GAD). Anxiety and Depression Association of America. Retrieved May 16, 2021, from https://adaa.org/understanding-anxiety/generalized-anxiety-disorder-gad

Anxiety Canada. (n.d.). Obsessive-Compulsive Disorder. Anxiety Canada. Retrieved May 16, 2021, from https://www.anxietycanada.com/disorders/obsessive-compulsive-disorder-2/

Applebaum, A. J., Kulikowski, J. R., & Breitbart, W. (2015). Meaning-Centered Psychotherapy for Cancer Caregivers (MCP-C): Rationale and Overview. Palliative & Supportive Care, 13(6), 1631–1641. https://doi.org/10.1017/S1478951515000450

Bonne, O., Grillon, C., Vythilingam, M., Neumeister, A., & Charney, D. S. (2004). Adaptive and maladaptive psychobiological responses to severe psychological stress: Implications for the discovery of novel pharmacotherapy. Neuroscience and Biobehavioral Reviews, 28(1), 65–94. https://doi.org/10.1016/j.neubiorev.2003.12.001

Borji, M., Nourmohammadi, H., Otaghi, M., Salimi, A. H., & Tarjoman, A. (2017). Positive Effects of Cognitive Behavioral Therapy on Depression, Anxiety and Stress of Family Caregivers of Patients with Prostate Cancer: A Randomized Clinical Trial. Asian Pacific Journal of Cancer Prevention : APJCP, 18(12), 3207–3212. https://doi.org/10.22034/APJCP.2017.18.12.3207

Breitbart, W. (2016). Meaning-Centered Psychotherapy in the Cancer Setting: Finding Meaning and Hope in the Face of Suffering. Oxford University Press.

Breitbart, W., & Duva, M. (2016). Meaning-Centered Psychotherapy in the Oncology and Palliative Care Settings (pp. 245–260). https://doi.org/10.1007/978-3-319-41397-6_12

Cuijpers, P., Quero, S., Dowrick, C., & Arroll, B. (2019). Psychological Treatment of Depression in Primary Care: Recent Developments. Current Psychiatry Reports, 21(12), 129. https://doi.org/10.1007/s11920-019-1117-x

Ehlers, A., Clark, D. M., Hackmann, A., McManus, F., & Fennell, M. (2005). Cognitive therapy for post-traumatic stress disorder: Development and evaluation. Behaviour Research and Therapy, 43(4), 413–431. https://doi.org/10.1016/j.brat.2004.03.006

Erskine, R. (2014). What Do You Say Before You Say Good-Bye? The Psychotherapy of Grief. Transactional Analysis Journal, 44, 279–290. https://doi.org/10.1177/0362153714556622

Fegg, M., Brandstätter, M., Kögler, M., Hauke, G., Rechenberg-Winter, P., Fensterer, V., Küchenhoff, H., Hentrich, M., Belka, C., & Borasio, G. (2013). Existential behavioural therapy for informal caregivers of palliative patients: A randomised controlled trial. Psycho-Oncology, 22. https://doi.org/10.1002/pon.3260

Foa, E. B. (2010). Cognitive behavioral therapy of obsessive-compulsive disorder. Dialogues in Clinical Neuroscience, 12(2), 199–207.

Foa, E. B., Yadin, E., & Lichner, T. K. (2012). Exposure and Response (Ritual) Prevention for Obsessive Compulsive Disorder: Therapist Guide (2nd ed.). Oxford University Press.

Frost, N., Laska, K., & Wampold, B. (2014). The Evidence for Present-Centered Therapy as a Treatment for Posttraumatic Stress Disorder. Journal of Traumatic Stress, 27, 1–8. https://doi.org/10.1002/jts.21881

Fulton, J. J., Newins, A. R., Porter, L. S., & Ramos, K. (2018). Psychotherapy Targeting Depression and Anxiety for Use in Palliative Care: A Meta-Analysis. Journal of Palliative Medicine, 21(7), 1024–1037. https://doi.org/10.1089/jpm.2017.0576

Gautam, M., Tripathi, A., Deshmukh, D., & Gaur, M. (2020). Cognitive Behavioral Therapy for Depression. Indian Journal of Psychiatry, 62(Suppl 2), S223–S229. https://doi.org/10.4103/psychiatry.IndianJPsychiatry_772_19

Government of Canada. (2016, October 18). Canadian Cancer Statistics [Notices]. Government of Canada. https://www.canada.ca/en/public-health/services/chronic-diseases/cancer/canadian-cancer-statistics.html

Gramm, J., Trachsel, M., & Berthold, D. (2020). Psychotherapeutic Work in Palliative Care. Verhaltenstherapie, 1–10. https://doi.org/10.1159/000505120

Hazlett-Stevens, H., & Craske, M. G. (2002). Brief Cognitive-Behavioral Therapy: Definition and Scientific Foundations. Wiley.

International OCD Foundation. (n.d.). What is OCD? International OCD Foundation. Retrieved May 15, 2021, from https://iocdf.org/about-ocd/

Klerman, G. L., & Weissman, M. M. (1994). Interpersonal Psychotherapy of Depression: A Brief, Focused, Specific Strategy. Jason Aronson, Incorporated.

Kosminsky, P. (2017). CBT for Grief: Clearing Cognitive Obstacles to Healing from Loss. Journal of Rational-Emotive & Cognitive-Behavior Therapy, 35. https://doi.org/10.1007/s10942-016-0241-3

Kristjanson, L. J., & Aoun, S. (2004). Palliative Care for Families: Remembering the Hidden Patients. The Canadian Journal of Psychiatry, 49(6), 359–365. https://doi.org/10.1177/070674370404900604

Markowitz, J. C., Petkova, E., Neria, Y., Van Meter, P. E., Zhao, Y., Hembree, E., Lovell, K., Biyanova, T., & Marshall, R. D. (2015). Is Exposure Necessary? A Randomized Clinical Trial of Interpersonal Psychotherapy for PTSD. The American Journal of Psychiatry, 172(5), 430–440. https://doi.org/10.1176/appi.ajp.2014.14070908

Neimeyer, R. A. (2012). Techniques of Grief Therapy: Creative Practices for Counseling the Bereaved. Routledge.

Nezu, A. M. (1986). Efficacy of a social problem-solving therapy approach for unipolar depression. Journal of Consulting and Clinical Psychology, 54(2), 196.

Radbruch, L., De Lima, L., Knaul, F., Wenk, R., Ali, Z., Bhatnaghar, S., Blanchard, C., Bruera, E., Buitrago, R., Burla, C., Callaway, M., Munyoro, E. C., Centeno, C., Cleary, J., Connor, S., Davaasuren, O., Downing, J., Foley, K., Goh, C., ... Pastrana, T. (2020). Redefining Palliative Care—A New Consensus-Based Definition. Journal of Pain and Symptom Management, 60(4), 754–764. https://doi.org/10.1016/j.jpainsymman.2020.04.027

Schnurr, P. P. (2017). Focusing on trauma-focused psychotherapy for posttraumatic stress disorder. Current Opinion in Psychology, 14, 56–60. https://doi.org/10.1016/j.copsyc.2016.11.005
Shear, M. K. (2012). Grief and mourning gone awry: Pathway and course of complicated grief. Dialogues in Clinical Neuroscience, 14(2), 119–128.

Shear, M. K., Ghesquiere, A., & Glickman, K. (2013). Bereavement and Complicated Grief. Current Psychiatry Reports, 15(11). https://doi.org/10.1007/s11920-013-0406-z

Sherman, A. C., & Simonton, S. (1999). Family Therapy for Cancer Patients: Clinical Issues and Interventions. The Family Journal, 7(1), 39–50. https://doi.org/10.1177/1066480799071006

Simon, N., Hollander, E., Rothbaum, B. O., & Stein, D. J. (2020). The American Psychiatric Association Publishing Textbook of Anxiety, Trauma, and OCD-Related Disorders. American Psychiatric Association Publishing.

Simos, G., & Hofmann, S. G. (2013). CBT For Anxiety Disorders: A Practitioner Book. John Wiley & Sons.

Vachon, M. L. (1988). Counselling and psychotherapy in palliative/hospice care: A review. Palliative Medicine, 2(1), 36–50. https://doi.org/10.1177/026921638800200107

Watkins, L. E., Sprang, K. R., & Rothbaum, B. O. (2018). Treating PTSD: A Review of Evidence-Based Psychotherapy Interventions. Frontiers in Behavioral Neuroscience, 12. https://doi.org/10.3389/fnbeh.2018.00258

Watson, M., & Kissane, D. W. (2011). Handbook of Psychotherapy in Cancer Care. John Wiley & Sons.

Wetherell, J. L. (2012). Complicated grief therapy as a new treatment approach. Dialogues in Clinical Neuroscience, 14(2), 159–166.

Wilson, J. P., Friedman, M. J., & Lindy, J. D. (2012). Treating Psychological Trauma and PTSD. Guilford Press.

Chapter Eight

Best Degree Programs. (n.d.). 5 Research Methods Used in Psychological Research. Best Degree Programs. https://www.bestdegreeprograms.org/lists/5-research-methods-used-in-psychology/

Dragioti, E., Karathanos, V., Gerdle, B., & Evangelou, E. (2017). Does psychotherapy work? An umbrella review of meta-analyses of randomized controlled trials. Acta Psychiatrica Scandinavica, 136(3), 236-246. https://doi-org.proxy.library.brocku.ca/10.1111/acps.12713

Fradera, A. (2017). Concerning study says psychotherapy has a problem with undeclared researcher bias. BPS. https://digest.bps.org.uk/2017/02/20/concerning-study-says-psychotherapy-research-has-a-problem-with-undeclared-researcher-bias/

Hunsley, J., Elliot, K., & Therrien, Z. (2013, September 10). The Efficacy and Effectiveness of Psychological Treatments. Canadian Psychological Association. https://cpa.ca/docs/File/Practice/TheEfficacyAndEffectivenessOfPsychologicalTreatments_web.pdf

Jarrett, C. (2017, March 20). Have we overestimated the effectiveness of psychotherapy? The British Psychological Society. https://digest.bps.org.uk/2017/03/20/have-we-overestimated-the-effectiveness-of-psychotherapy/

McNeilly, C. L., & Howard, K. I. (2008). The effects of psychotherapy: A re-evaluation based on dosage. Psychotherapy Research, 1(1), 74-78. https://www.tandfonline.com/doi/abs/10.1080/10503309112331334081#:~:text=In%201952%2C%20Hans%20Eysenck%20published,than%20that%20of%20spontaneous%20remission.&text=Thus%2C%20Eysenck's%20data%20reveal%20that%20psychotherapy%20is%20very%20effective.

Munder, T., Flückiger, C., Leichsenring, F., Abbass, A. A., Hilsenroth, M. J., Luyten, P., Rabung, S., Steinert, C., & Wampold, B. E. (2018). Is psychotherapy effective? A re-analysis of treatments for depression. Epidemiology and Psychiatric Sciences, 28(3). https://www.cambridge.org/core/journals/epidemiology-and-psychiatric-sciences/article/is-psychotherapy-effective-a-reanalysis-of-treatments-for-depression/5D8EC85B6FA35B5CEE124381F18E51B9#article

CHAPTER NINE
Altman, M., Jacobi, L., Avilla, R., Beston, B., Brown, K., Burton, E. T., ... Wehe, H. (2017). Introduction to psychology: A Top Hat interactive text. Top Hat Monocle. https://tophat.com/marketplace/beta/oer-introduction-to-psychology-meaghan-altman/737/

Bishop, F. M. (n.d.). How Is Modern Psychotherapy Different? Psychology Today. https://www.psychologytoday.com/ca/blog/managing-addictions/202010/how-is-modern-psychotherapy-different.

Pajer, N. (2019, September 9). This Is What Therapy Is Like Around The World. HuffPost. https://www.huffpost.com/entry/what-therapy-is-like-around-the-world_l_5d2f2930e4b02fd71dddab1c.

Carey, B. (2004). For psychotherapy's claims, skeptics demand proof. The New York Times. https://www.nytimes.com/2004/08/10/science/for-psychotherapy-s-claims-skeptics-demand-proof.htmHunt, J. (n.d.). The dangers of holding therapy. The Natural Child Project. https://www.naturalchild.org/articles/jan_hunt/holding_therapy.html#:~:text=Holding%20therapy%20is%20a%20practice,until%20there%20is%20eye%20contact

Cook, S. C., Schwartz, A. C., & Kaslow, N. J. (2017). Evidence-based psychotherapy: Advantages and challenges. Neurotherapeutics, 14(3), 537-545. https://doi.org/10.1007/s13311-017-0549-4

Holmes, J.D., & Beins, B.C. (2009). Psychology is a science: At least some students think so. Teaching of Psychology, 36, 5-11. doi:10.1080/00986280802529350

Hong, D. S. (2015). Here/In this issue and there/Abstract thinking: The art (and science) of psychotherapy. Journal of the American Academy of Child & Adolescent Psychiatry, 54(3), 157-158. https://doi.org/10.1016/j.jaac.2014.12.013

Hupp, S., Mercer, J., Thyer, B. A., & Pignotti, M. (2019). Critical thinking about psychotherapy. Pseudoscience in Child and Adolescent Psychotherapy, 1-13. https://doi.org/10.1017/9781316798096.003

Lilienfeld, S. O., & Gurung, R. A. (2012). Tackling student skepticism of psychology: Recommendations for instructors. https://www.apa.org. https://www.apa.org/ed/precollege/ptn/2012/08/student-skepticism

Meier, A. (n.d.). Emotional disturbance. Montana.gov. https://dphhs.mt.gov/schoolhealth/chronichealth/developmentaldisabilities/emotionaldisturbance#:~:text=Emotional%20disturbance%20means%20a%20condition,%2C%20sensory%2C%20or%20health%20factors/

CHAPTER ELEVEN

Andrews, G., Issakidis, C., & Carter, G. (2001). Shortfall in mental health service utilization. British Journal of Psychiatry, 179, 417–425.

Angermeyer, M. C., van der Auwera, S., Carta, M. G., & Schomerus, G. (2017). Public attitudes towards psychiatry and psychiatric treatment at the beginning of the 21st century: a systematic review and meta-analysis of population surveys. World psychiatry : official journal of the World Psychiatric Association (WPA), 16(1), 50–61. https://doi.org/10.1002/wps.20383

Beker, A. (2015, April 07). Public perception of counseling and its meaning in society. Retrieved May 12, 2021, from https://austinfamilycounseling.com/public-perception-of-counseling/

Ben-Porath, D.D. (2002). Stigmatization of individuals who receive psychotherapy: An interaction between help-seekingbehavior and the presence of depression. Journal of Social & Clinical Psychology, 21, 400-413.

Brym, R. J. (2019). SOC+.

Center for Mental Health Services. (2000). Anti-stigma: Do you know the facts? Rockville, MD: Author. Retrieved May 12, 2021, from http://www.mentalhealth.org/stigma/factsheet.htm

Corrigan, P.W. (2004). How stigma interferes with mental health care. American Psychologist, 59, 614-625.

Deane, F.P., & Chamberlain, K. (1994). Treatment fearfulness and distress as predictors of professional psychological help-seeking. British Journal of Guidance and Counseling, 22, 207-217.

Deane, F.P., & Todd, D.M. (1996). Attitudes and intentions to seek professional psychological help for personal problems or suicidal thinking. Journal of College Student Psychotherapy,10 (4), 45-59.

Gaudiano, B. A. (2013, September 30). Psychotherapy's image problem. Retrieved May 12, 2021, from https://www.nytimes.com/2013/09/30/opinion/psychotherapys-image-problem.html

Hill, C. E., Satterwhite, D. B., Larrimore, M. L., Mann, A. R., Johnson, V. C., Simon, R. E., Simpson, A. C., & Knox, S. (2012). Attitudes about psychotherapy: A qualitative study of introductory psychology students who have never been in psychotherapy and the influence of attachment style. Counselling & Psychotherapy Research, 12(1), 13–24. https://doi.org/10.1080/14733145.2011.629732

Hinson, J.A., & Swanson, J.L. (1993). Willingness to seek help as a function of self-disclosure and problem severity. Journal of Counseling & Development, 71, 465-470.

Jorm, A. F., Christensen, H., & Griffiths, K. M. (2005). The impact of beyondblue: the national depression initiative on the Australian public's recognition of depression and beliefs about treatments. The Australian and New Zealand journal of psychiatry, 39(4), 248–254. https://doi.org/10.1080/j.1440-1614.2005.01561.x

Kohli, S. (2016, May 20). 14 common misconceptions about people who go to therapy. Retrieved May 12, 2021, from https://www.huffpost.com/entry/misconceptions-about-therapy_b_7286204

Layton, J. (2020, June 30). How exorcism works. Retrieved May 12, 2021, from https://science.howstuffworks.com/science-vs-myth/afterlife/exorcism.htm

Lumen Learning. (n.d.). Introduction to psychology. Retrieved May 12, 2021, from https://courses.lumenlearning.com/wsu-sandbox/chapter/mental-health-treatment-past-and-present/

Leech, N.L., (2007). Cramer's model of willingness to seek counseling: A structural equation model for counseling students. The Journal of Psychology, 141, 435-445

Nemec, K. (2005). Public health depression initiative: A review of depression cam-paigns. Lessons for New Zealand (R4111-LR-3.doc.doc). Retrieved May 12, 2021, from http://www.moh.govt.nz/moh.nsf/indexmh/nationaldepression-initiative

Philo, G. (1994). Media representations of mental health/illness: Audience reception study. Glasgow: Glasgow University Media Group.

Potter, W. J., & Chang, I. K. (1990). Television exposure measures and the cultivation hypothesis. Journal of Applied Communication Research, 27, 258–284.

Ryan, M., Ryan, P., & Ryan Rosenbaum, K. (2017, July 16). Shifts in public perception of the therapeutic process • duffy counseling center. Retrieved May 12, 2021, from https://duffycounseling.com/2017/07/16/shifts-in-public-perception-of-therapy/

Shaffer, P.A., Vogel, D.L., & Wei, M. (2006). The mediating roles of anticipated risks, anticipated benefits, and attitudes on the decision to seek professional help: An attachment perspective. Journal of Counseling Psychology, 53, 442-452

Snyder, J.F., Hill, C.E., & Derksen, T.P. (1972). Why some students do not use university counseling facilities. Journal of Counseling Psychology, 19, 263-268)

Szasz, T. S. (1960). The myth of mental illness. American Psychologist, 15(2), 113–118. https://doi.org/10.1037/h0046535

CHAPTER TWELVE
Aftab, A. (2020). Integration, Common Principles, and the Future of Psychotherapy: Marvin R. Goldfried, PhD. Psychiatric Times. https://www.psychiatrictimes.com/view/integration-common-principles-future-psychotherapy

Essig, T. (2019). The War For The Future Of Psychotherapy. Forbes. https://www.forbes.com/sites/toddessig/2019/12/27/the-war-for-the-future-of-psychotherapy/?sh=28860152759b

Fiske, A., Henningsen, P., Buyx, A. (2019). Your Robot Therapist Will See You Now: Ethical Implications of Embodied Artificial Intelligence in Psychiatry, Psychology, and Psychotherapy. J Med Internet Res, 21(5). DOI: 10.2196/13216

Huberty, J., Green, J., Glissmann, C. (2019). Efficacy of the Mindfulness Meditation Mobile App "Calm" to Reduce Stress Among College Students: Randomized Controlled Trial. JMIR Mhealth Uhealth, 7(6). DOI: 10.2196/14273

McDonald, A. et al (2020). Online psychotherapy: Trailblazing digital healthcare. BJPsych Bulletin, 44(2), 60-66. DOI :10.1192/bjb.2019.66

Pfefferbaum, B., North, C.S. (2020) Mental Health and the Covid-19 Pandemic. N Engl J Med, 383,510-512. DOI: 10.1056/NEJMp2008017

Understanding psychotherapy and how it works. (2012). American Psychological Association. https://www.apa.org/topics/psychotherapy/understanding

What is Psychotherapy? (2019). American Psychiatric Association. https://www.psychiatry.org/patients-families/psychotherapy

Yang, R., Vigod, S.N., Hensel, J.M. (2019). Optional Web-Based Videoconferencing Added to Office-Based Care for Women Receiving Psychotherapy During the Postpartum Period: Pilot Randomized Controlled Trial. J Med Internet Res, 21(6). DOI: 10.2196/13172

www.ingramcontent.com/pod-product-compliance
Lightning Source LLC
Chambersburg PA
CBHW030118170426
43198CB00009B/654